I0036945

ISBN: 9780648608752

To Georgia and Scarlett,

Watching you grow is like a movie that changes each day. Some days it's a comedy and other days more like Fight Club. It's the Pitch Perfect days that I fear most.

Children are remarkable for their ability to quickly and easily learn massive amounts of new information. It is proposed from recent research that the key behind this ability to rapidly process is the neurotransmitter GABA. Furthermore, functional MRI have proven children activate different and more regions of their brains than adults when they perform word tasks*.

*Washington University School of Medicine in St. Louis.

Why is it then if children's brains absorb information at a much faster rate than adults that they are so hard to teach at home? We have a theory...

"The average 4-year-old laughs 300 times a day. The average 40-year-old? Only four."

This common quote is hard to attribute to a single individual and trace its origin. However It is undeniable to a parent that children laugh longer and louder and are more distracted than adults.

The desire to have fun competes with their increased capability to learn. Imagine a world where you both win. They do their homework (tick parenting win) and they have fun (tick child win). This lead to our research and development of Learning through fun.

This series focuses on mathematics and is based upon the New Zealand Curriculum, Australian Schools Curriculum (Version 9) and Naplan testing 2022. Each book contains a series of fun Dot to Dot pictures, joke puzzles and colour by number activities. This book introduces MAB building blocks as part of the maths curriculum.

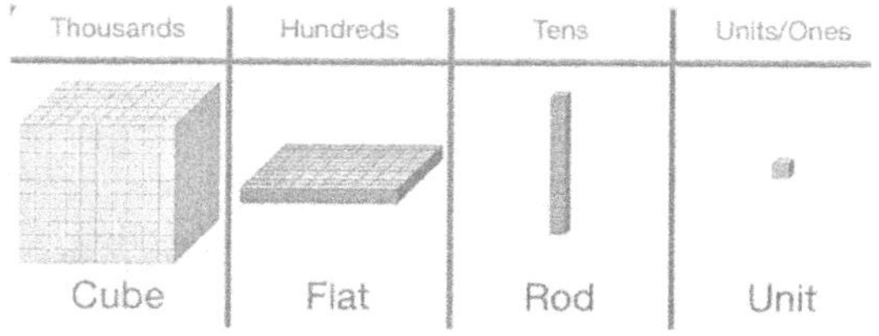

- Index -

PUZZLE 1

Solve the equations and connect the dots in order

MATHS PROBLEM	ANSWER
Start:	
25 x 2	
1/2 of 50	
100 ÷ ? = 50	
40 - 13	
2 x 5	
100 - 51	
12 + ? = 50	
250 - 207	
3 x 3	
7 x 10	
40 - 11	
3 x 5	
400÷ 10	
6 x 8	
5 x 6	
62-4	
100 - 15	
3 x ? =21	
20 +30 + 23	
10 = 41 - ?	
5 x ? = 100	
30 -13	
35 +36	

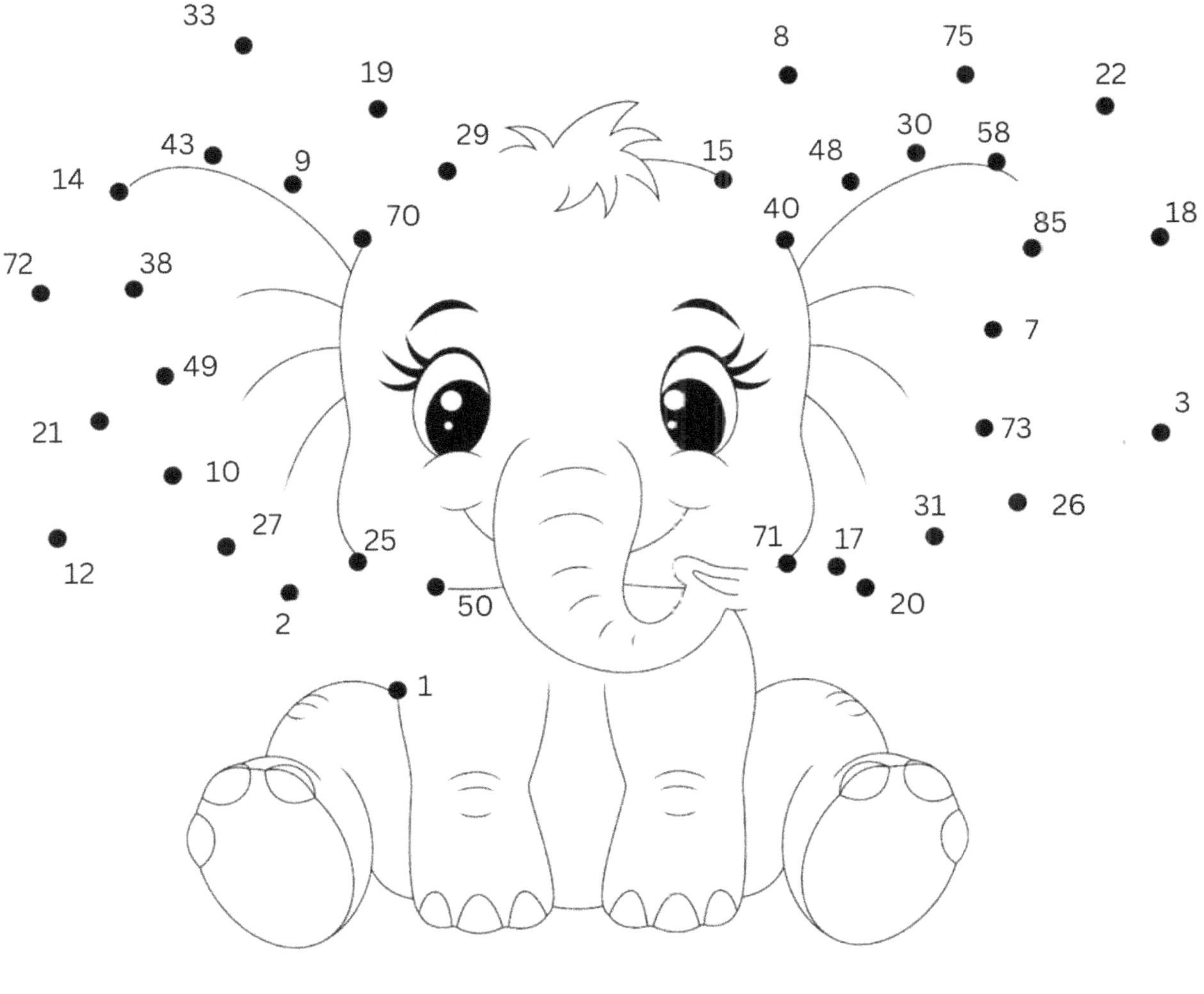
33
8
75
19
22
43
29
30
58
14
9
15
48
70
40
85
18
72
38
7
49
3
21
73
10
31
26
27
25
71
17
12
50
20
2
1

PUZZLE 2

Solve the equations and connect the dots in order

MATHS PROBLEM	ANSWER
5, ?, 25, 35	
7 x 11	
2 is 1/4 of ?	
4 x4	
50-13	
30 x 3	
60 -6	
2 x 5	
21, 26, ?, 36	
100 -20	
50 - 33	
28 - 21	
16 + 16	
30 = ? - 61	
41 + 30	
26 = ? - 13	
? x 9 = 27	
6 x 11	
100 -24	
12 + 22	
5 x 12	

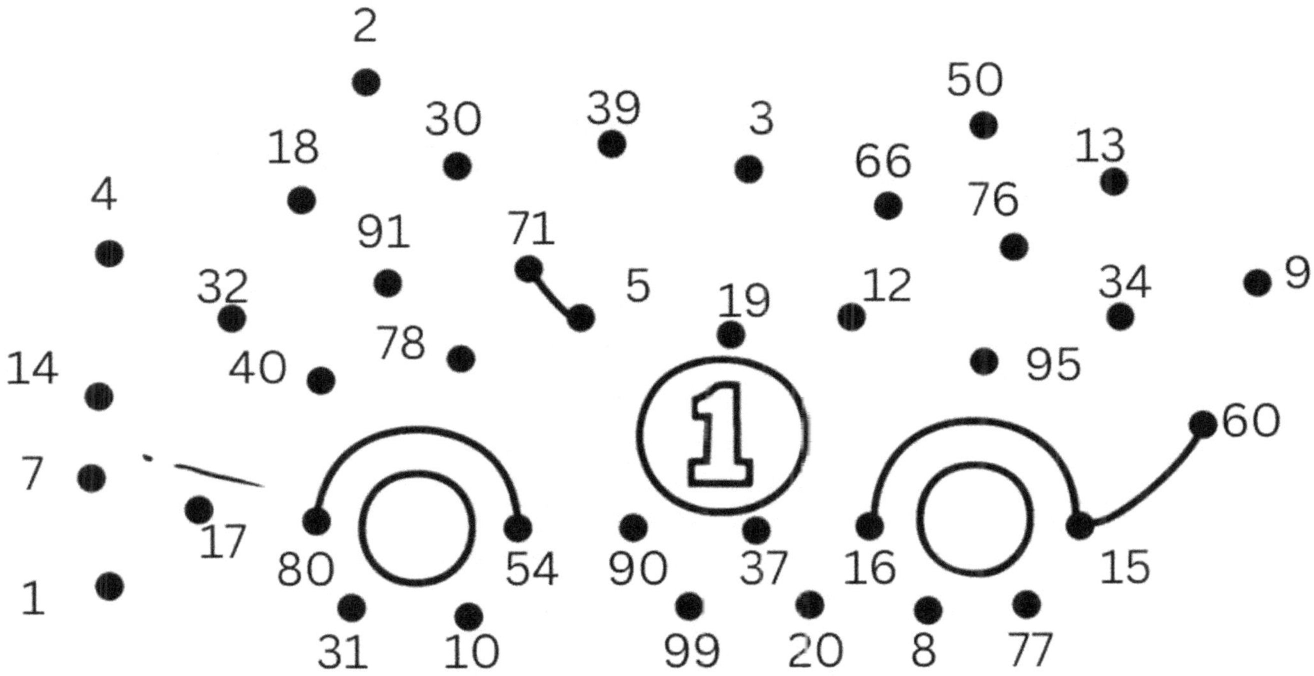
2
50
30
39
3
18
66
13
4
76
91
71
9
32
5
19
12
34
78
14
40
95
60
7
17
80
54
90
37
16
15
1
31
10
99
20
8
77

PUZZLE 3

Solve the equations and connect the dots in order

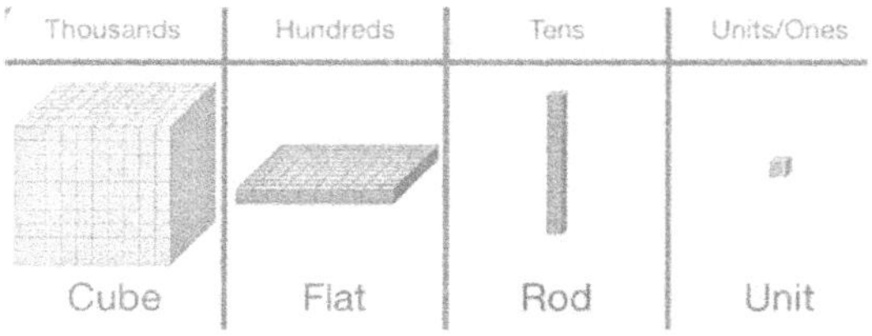

MATHS PROBLEM	ANSWER
5 x 12	
60 -9	
30 -11	
25 + 12	
4  3	
3  2  1 	
6 x 11	
44 - 11	
2 x 11	
90 - 13	
21 -14	
4 x 5	
189 - 188	
11 x 5	
I have four 100's, two 10's and three singles	
11 x 9	
3  4	
82 - 2	
3  6 	

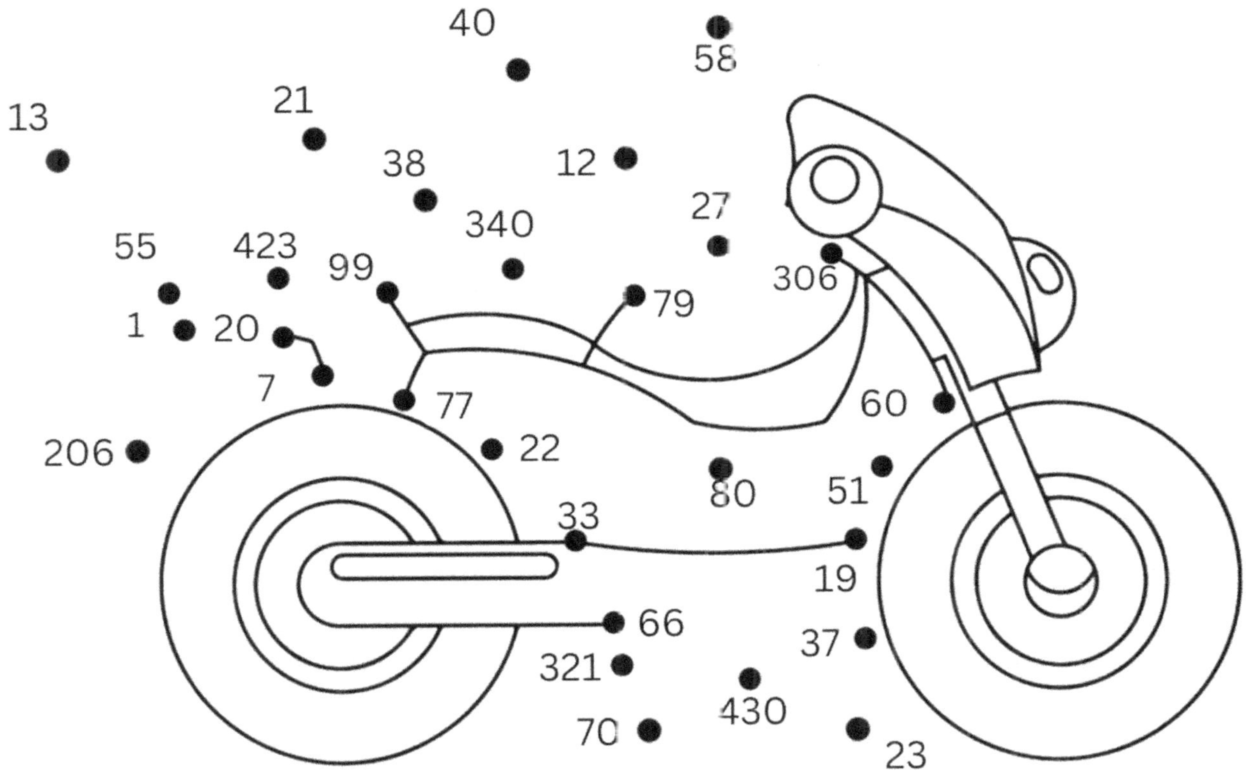

40
58
21
13
38
12
27
340
55
423
306
99
79
1
20
7
77
60
206
22
80
51
33
19
66
37
321
430
70
23

PUZZLE 4

Solve the equations and connect the dots in order

MATHS PROBLEM	ANSWER
start: seventy minus nine	
20 + 10 + 17	
2  9	
100 - 49	
44- 4	
100 - 31	
5 x 11	
33 ÷ 11	
3 x 11	
40 + 13	
four times eleven	
3 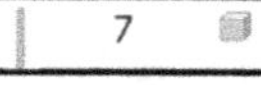 7	
60 - 4	
5 x 5	
80 - 9	
100 - 22	
5 x 12	
3  4	
5 x ? = 35	
10 + 10 + 9	
half of sixty	
8 x 11	
2  5	
? ÷ 3 = 30	
seven times ten	
20 + 20 + 1	
40 -1	
2 x 11	
24 ÷ ? = 12	
seventy minus eight	
100 - 9	
7 x 11	
6 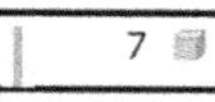 7	
40 - ? = 9	
7 x 6	
109 - ? = 50	
6 x 12	

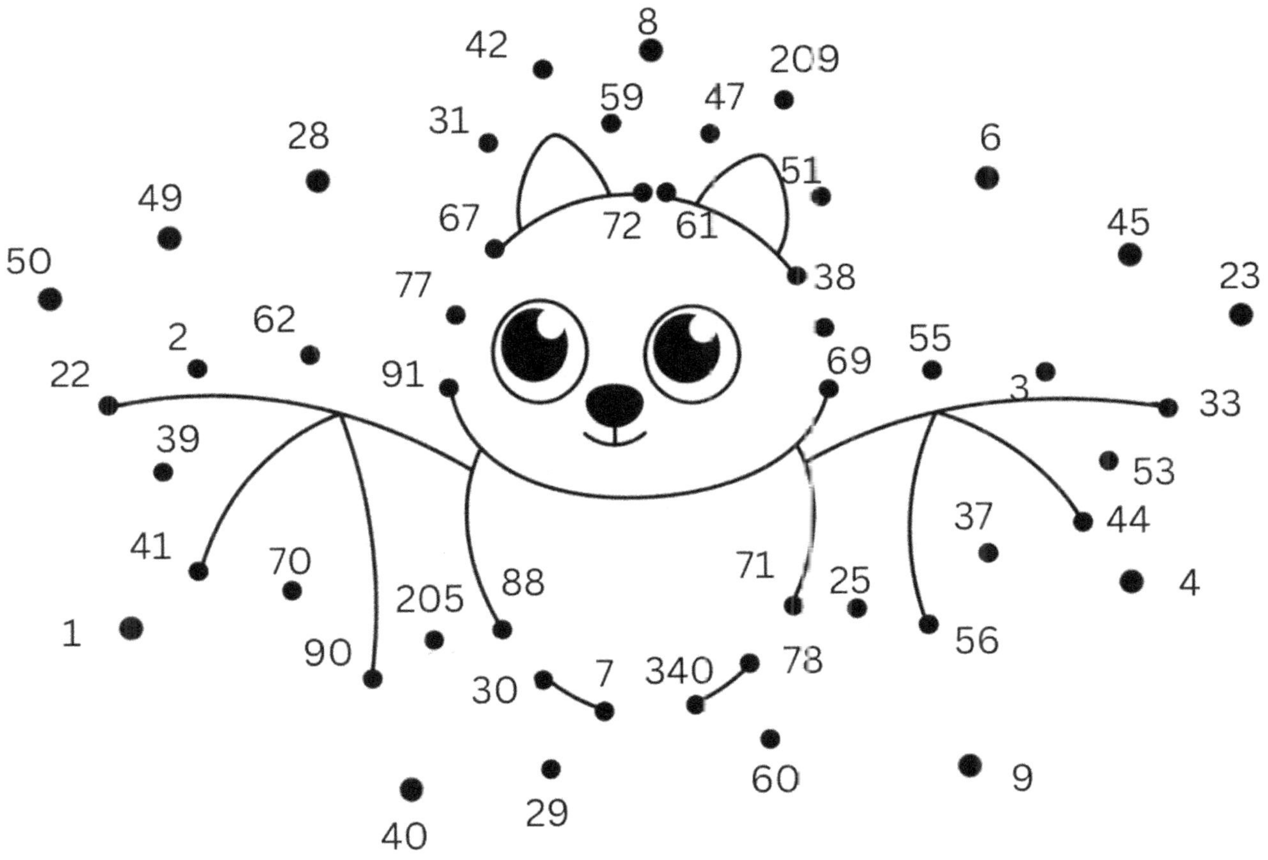
8
42
209
59
47
31
28
6
51
49
67
72
61
45
50
23
38
77
62
2
55
22
91
69
3
33
39
53
37
44
41
71
70
25
4
88
205
1
90
78
7
340
56
30
60
9
40
29

PUZZLE 5

A Joke for you

LETTER	ANSWER
A	3
B	4
C	7
D	9
E	12
F	14
G	16
H	18
I	24
J	27
K	30
L	31
M	33
N	35
O	38
P	40
Q	41
R	44
S	45
T	48
U	50
V	55
W	66
X	72
Y	77
Z	84

Why did the banana go the the doctor?

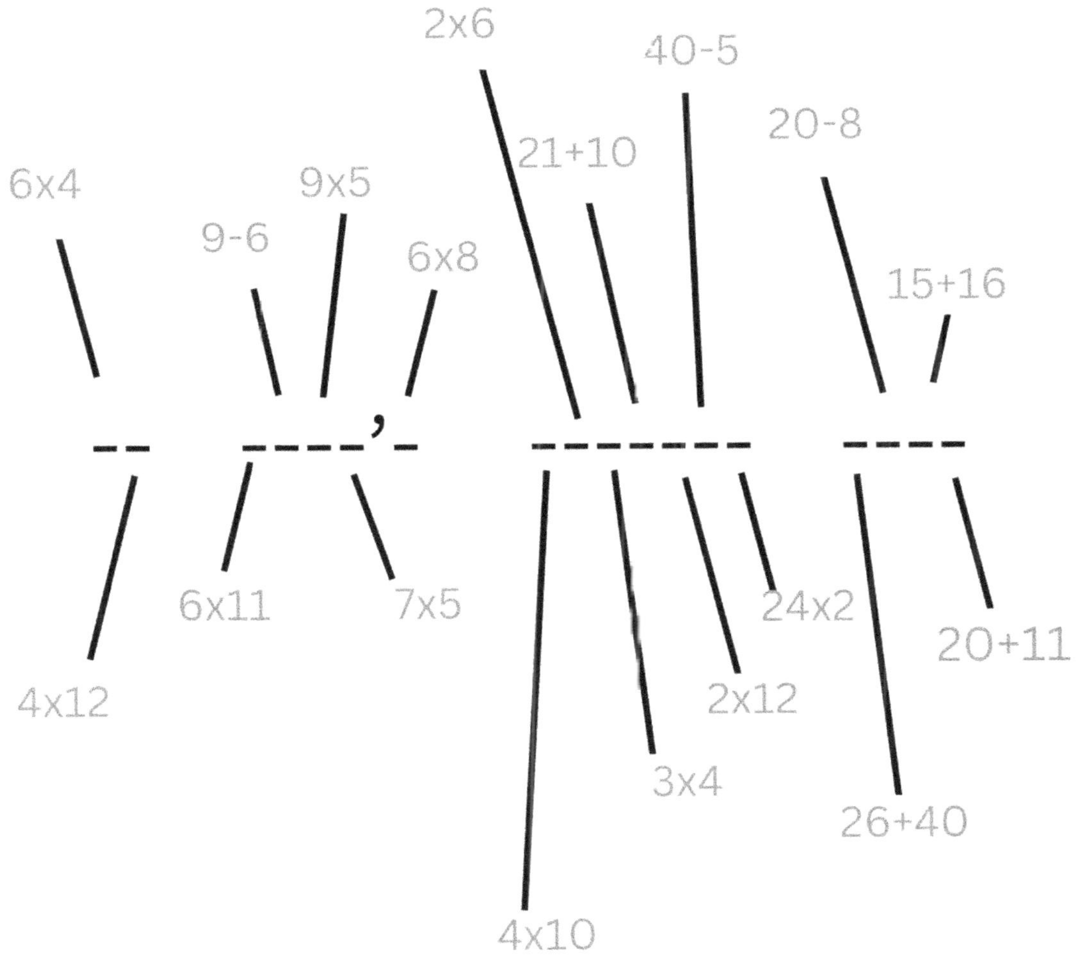

PUZZLE 6

Solve the equations and connect the dots in order

MATHS PROBLEM	ANSWER
start: 60 - 42 = 17 + ?	
10 + 10 + 10 + 10	
29 - 20	
20 x 3	
7 x 2	
11 x 7	
3 x 9	
3  3 	
48 ÷ ? = 12	
3 x 4	
3  2 	
29 + ? = 60	
20 + 37	
3 X 6	
60 -19	
85 - 6	
? X 5 = 35	
13 X 2	
2  3  4 	
40 - 12	
8 + 11	
6 x 4	
40 + 3	
2  6 	
26 + 10 + 20	
3 x 11	
3 x 3	
2 x 11	
4 x 5	
4 x 11	
1/2 of 100	
41 + 40	
41 + 20	

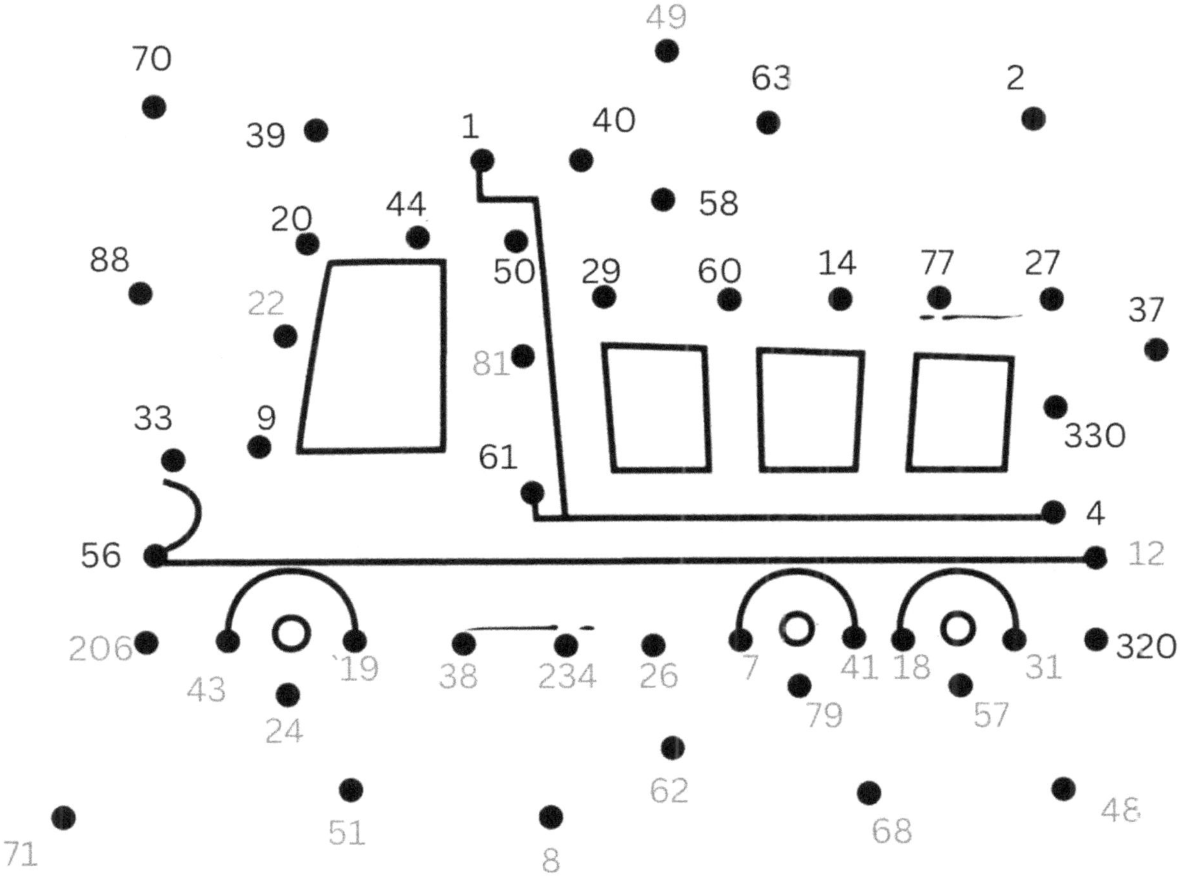

49
70
63
2
39
1
40
58
20
44
88
50
29
60
14
77
27
22
37
81
33
9
330
61
4
56
12
206
43
`19
38
234
26
7
41
18
31
320
24
79
57
62
51
8
68
48
71

PUZZLE 7

Colour by Number

MATHS PROBLEM	ANSWER
BLACK	1, 5, 10, 12
RED	13, 26, 50
GREY	2, 9, 14
YELLOW	7, 33, 100
LIGHT BLUE	3, 8, 24
ORANGE	6, 11,15
WHITE	4, 16 20, 70

2X2
8+8
2+2
18-14
20 + 50
4x5
4+12
9+7
2X5
25-24
2 X 2
2 X 2
2 X 10
6-5
10-9
10 +10
5+15
2+1
5-2
10+10
13-9
6X2
20 + 30
25-24
9+7
10
+ 6
9-4
30+40
4x5
2X2
65+5
17-16
17-13
100-90
14
-10
6-2
2X2
17
-1
2X5
30
- 20
18-14
30+40
2X10
4X3
10-6
2X2
4x4
4+12
5 X 2
4-3
8+8
2+8+10
7+9
4+12
2X5
6-1
12-7
10-5
10
+10
10+5+5
10+6
3 +9
8+8
18-2
34-30
17-1
3+13
17-1
2X2
50 X 2
13+20
20 +30+50
10-3
2 X 8
100 -30
17-1
2X2
2 X 8
2X2

PUZZLE 8

Colour by Number

MATHS PROBLEM	ANSWER
BLACK	11
RED	5
GREY	3
LIGHT BLUE	4
YELLOW	2
ORANGE	6

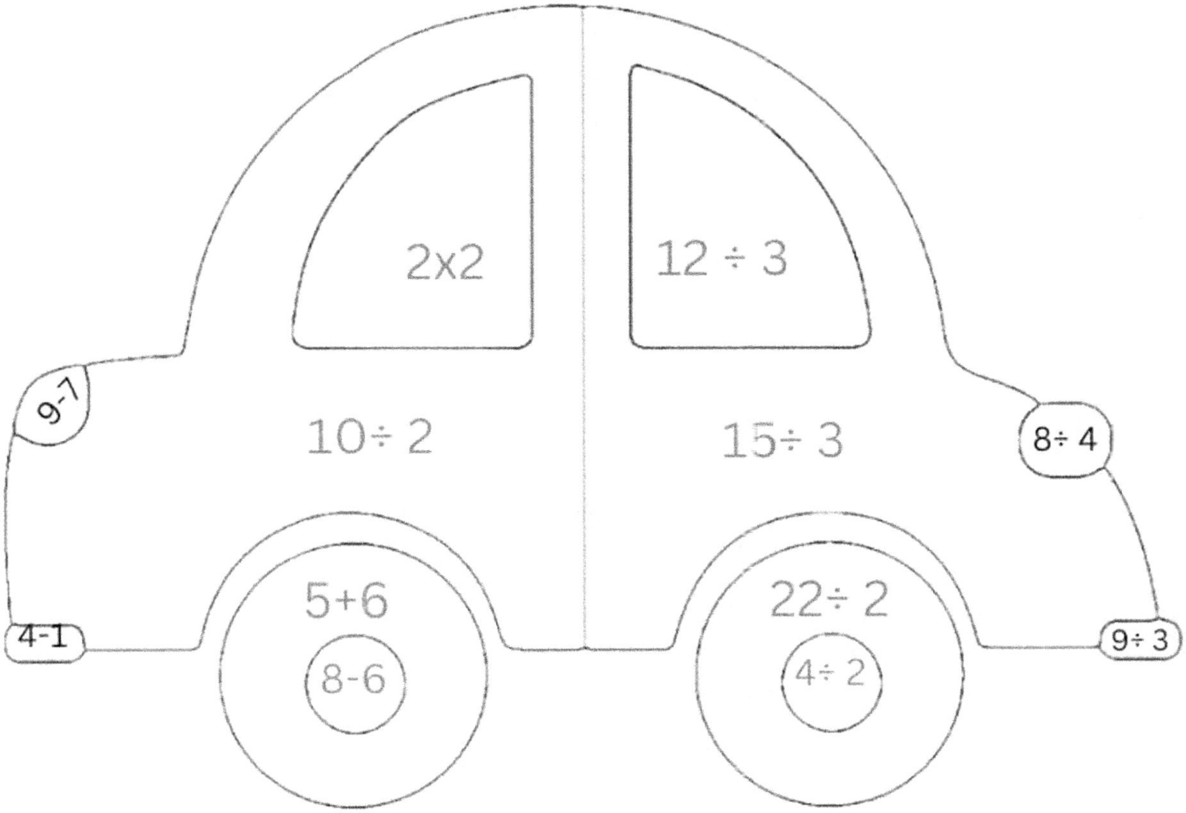
2x2
12 ÷ 3
9-7
10÷ 2
15÷ 3
8÷ 4
5+6
22÷ 2
4-1
9÷ 3
8-6
4÷ 2

PUZZLE 9

Solve the equations and connect the dots in order

MATHS PROBLEM	ANSWER
29 + 10	
87 - ? = 86	
5 x 8	
5 x 5	
73 - 20	
4 x 4	
2 x 4	
4 2	
6 x 4	
2 2 1	
21 + 30	
3 x 3	
9 + 10	
6 9	
9 x 11	
6 3 5	
3 x 4	
? x 12 = 60	
4 x 12	
12 ÷ ? = 3	
100 -10	
6 x 7	
20 x 4	
22 x 4	
5 x 12	
14 + 23	
44 ÷ 2	
? x 5 = 30	
6 x 12	
44 ÷ ? =4	
5 x 11	
8 x ? = 24	
2 5	
5 x 4	
11 + 10 + 20	
18 ÷ ? = 9	

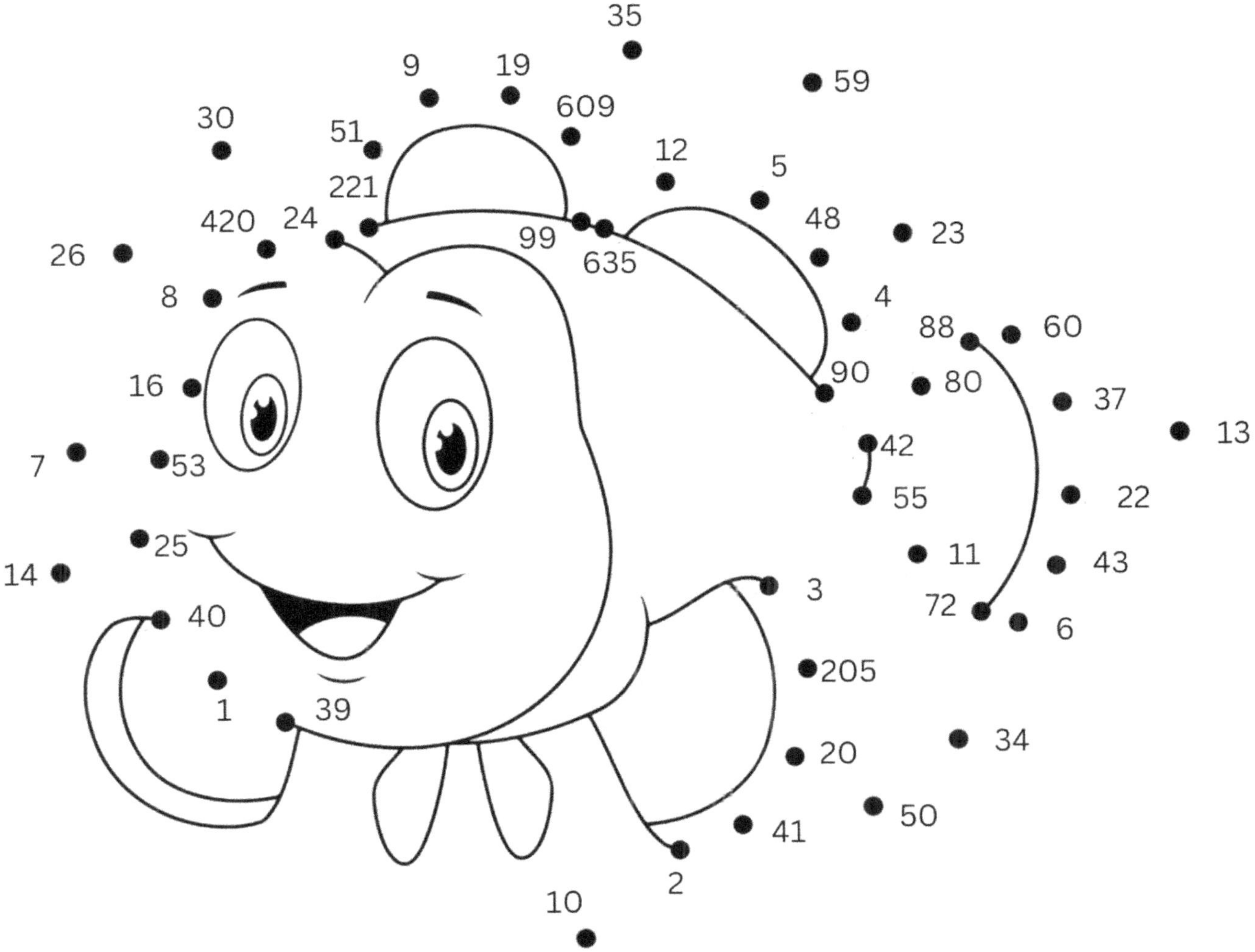
35
9
19
59
609
30
51
12
5
221
48
420
24
99
23
26
635
8
4
88
60
16
90
80
37
13
42
7
53
55
22
25
11
43
14
3
72
40
6
205
1
39
34
20
50
41
2
10

PUZZLE 10

Solve the equations and connect the dots in order

MATHS PROBLEM	ANSWER
start: 21 + 40	
100 -18	
25 x 4	
3  4	
4 x 11	
50 - ? = 1	
29 + 10	
30 - 13	
30 x 3	
1  3 6 	
3 x 10	
3 x 5	
4 x 20	
5 x 9	
2  1 2	
6 x 3	
10 x 4	
5 x 5	
35 x 2	
4  6	
50 - 9	
16 x 2	
5 x 2	
100 - 39	

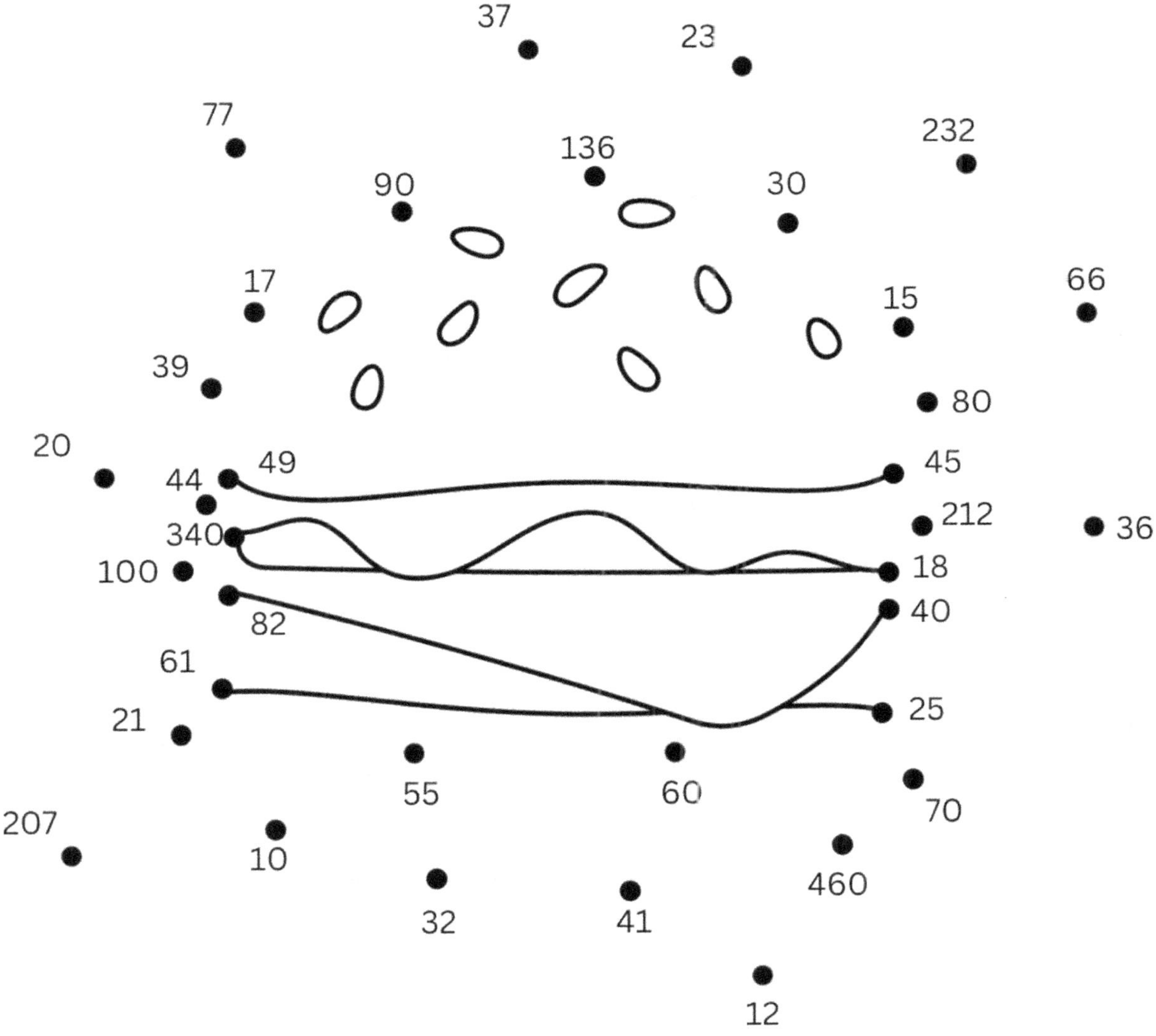

37
23
77
136
232
90
30
17
15
66
39
80
20
44
49
45
340
212
36
100
18
82
40
61
25
21
55
60
70
207
10
460
32
41
12

PUZZLE 11

Solve the equations and connect the dots in order

MATHS PROBLEM	ANSWER
1/4 of 40	
40 - 11	
60 - 8	
6  3 	
6 x 12	
11 x 8	
60 -1	
6 x 4	
3  2 7 	
5 X 9	
25 +26	
8 + 11	
30 x 2	
half of 30	
3 x 30	
? ÷ 3 = 9	
9 x 11	
3 x 11	
8 x 5	
50 + 32	
5  2 4 	
2 x 11	
10 + 3	
2  3 1 	

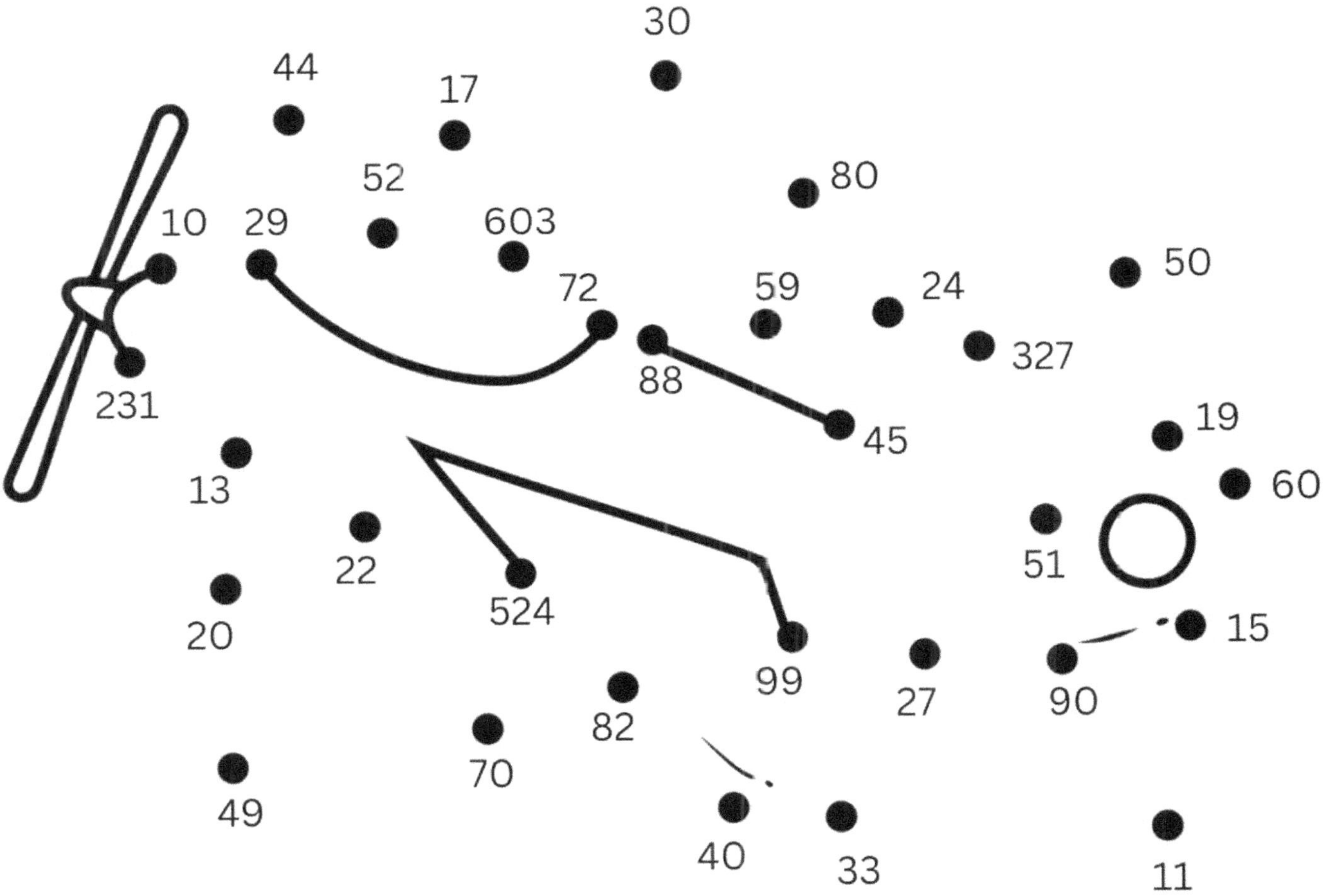
30
44
17
52
80
10
29
603
50
72
59
24
327
88
231
19
45
60
13
22
51
524
20
15
99
27
90
82
70
49
40
33
11

PUZZLE 12

A Joke for you

LETTER	ANSWER
A	3
B	4
C	7
D	9
E	12
F	14
G	16
H	18
I	24
J	27
K	30
L	31
M	33
N	35
O	38
P	40
Q	41
R	44
S	45
T	48
U	50
V	55
W	66
X	72
Y	77
Z	84

What is a cow's favourite cereal?

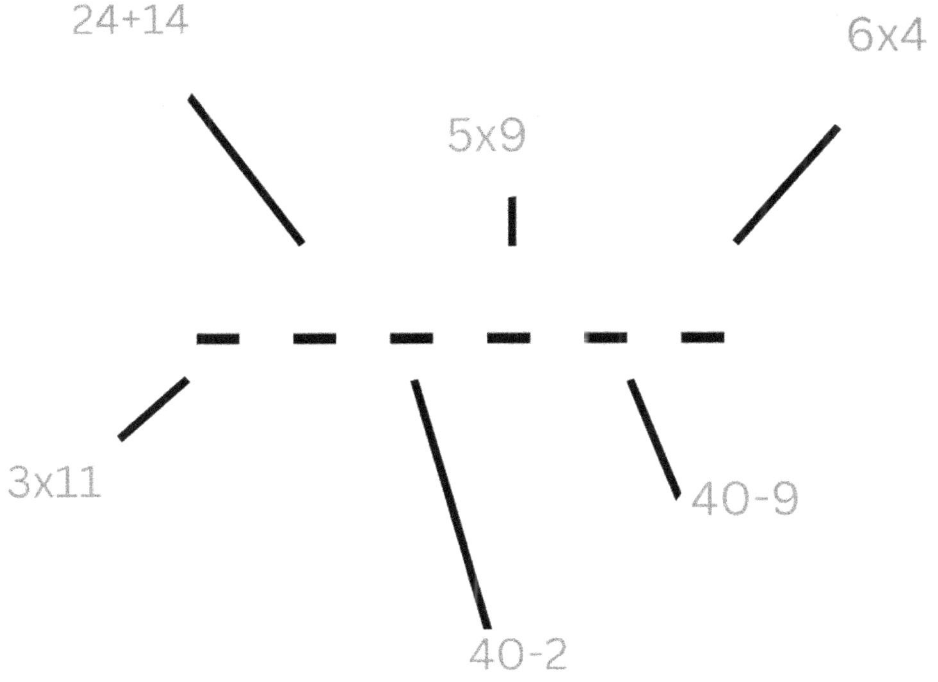

PUZZLE 13

Colour by Number

MATHS PROBLEM	ANSWER
BLACK	1, 7, 42
GREEN	5, 11, 24
GREY	3, 8, 32
LIGHT BLUE	4, 27, 50
YELLOW	2, 9, 44
ORANGE	6, 23, 56
WHITE	10,14, 25

60÷6
30-5
5x2
20-6
8-6
11-9
100÷10
5X5
20-9
5+6
16-14
24-10
4x6
35÷7
6-4
9-4
33÷3
2x12
7X2
7-5
40÷4
8-6
6x4
34-10
2+8
5-3
3x8
15÷3
28÷2
5+9
32-7
90-79
2+8
5X5
13-2
22÷2
50÷2
20÷4
25÷5
8-3
4+6
5x2
40÷8
16+9
20÷2
44-34
2x5
2X7

PUZZLE 14

Solve the equations and connect the dots in order

MATHS PROBLEM	ANSWER
Start: 46 = 26 + 19 + ?	
30 x 2	
5 x3	
? ÷ 3 = 10	
22 + 30	
30 -13	
9 x 11	
20 + 50	
3 x 6	
32 + 27	
33 - 22	
33 + 36	
100 - 9	
25 + 36	
33 + 20	
22 - 2	
? x 5 = 30	
7 x 10	
6 x 12	
108 - 16	
47 - 33	
126 - 40	
9 x ? = 45	
100 -19	
80 ÷ ? = 8	
5 + 8	
100 - 49	
70 - 35	
2 x 10	

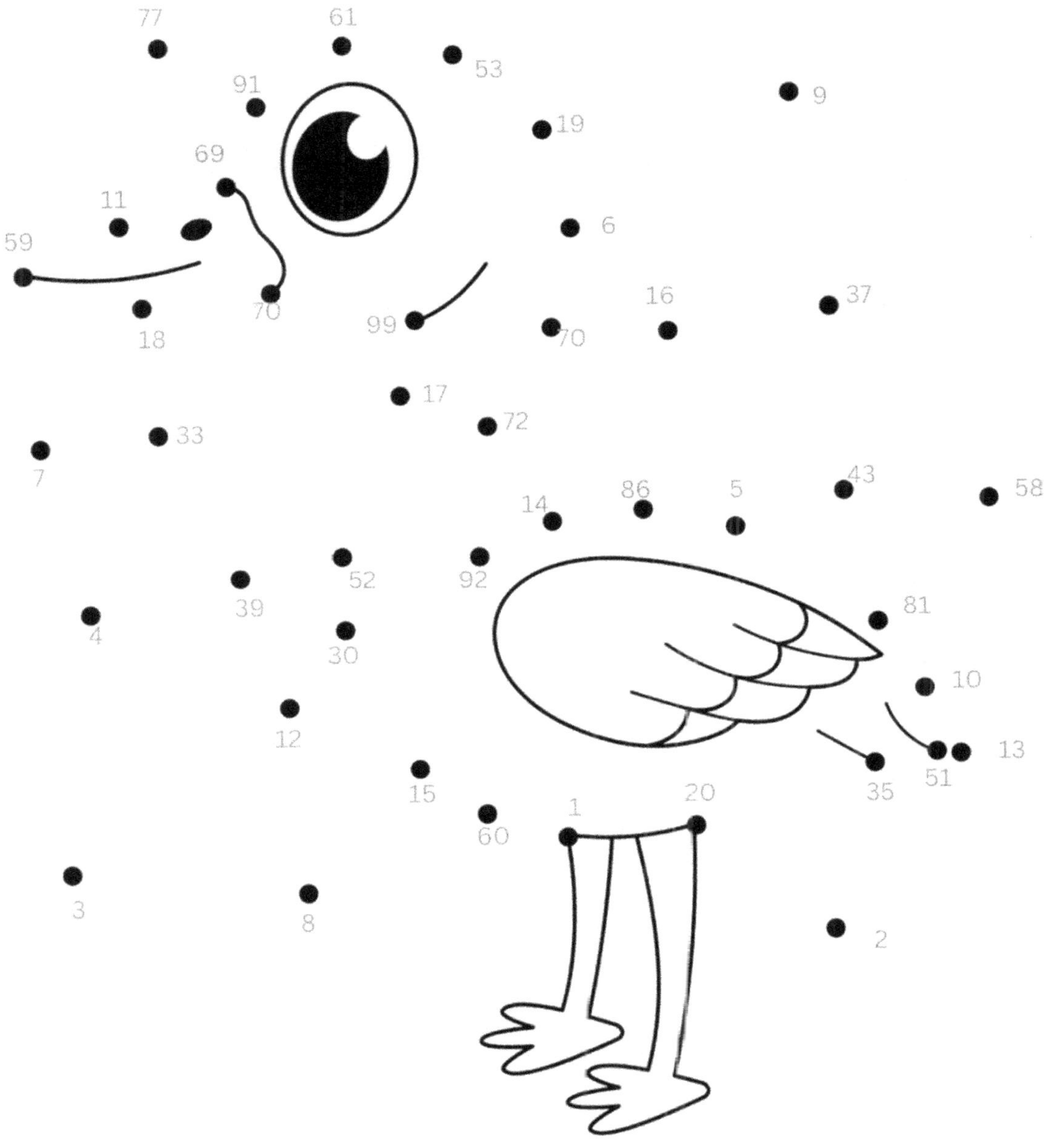
77
61
53
91
9
19
69
11
6
59
70
18
99
70
16
37
17
72
33
7
43
14
86
5
58
52
92
39
81
4
30
10
12
35
51
13
15
1
20
60
3
8
2

PUZZLE 15

Solve the equations and connect the dots in order

MATHS PROBLEM	ANSWER
start: 46 - ? = 9 x 5	
42 + 50	
9 x 3	
72 ÷ 12	
? x 2 = 26	
50 - 13	
3 x 6	
35 ÷ ? = 5	
1/4 of 100	
? - 20 = 21	
100 - 26	
23 + 26	
104 - 10	
4 x 4	
56 - 43	
46 + 30	
14 + 20	
21 + 40	
9 x 5	
84 - 6	
11 x 2	
24 ÷ 12	
5 x 12	
3 x ? = 27	
100 - 4	
2 x 7	
11 x 5	
75 - 6	
20 + 30 + 29	
? ÷ 7 = 3	
104 - 6	
65 -14	
20 + 20 + 7	
400 -399	

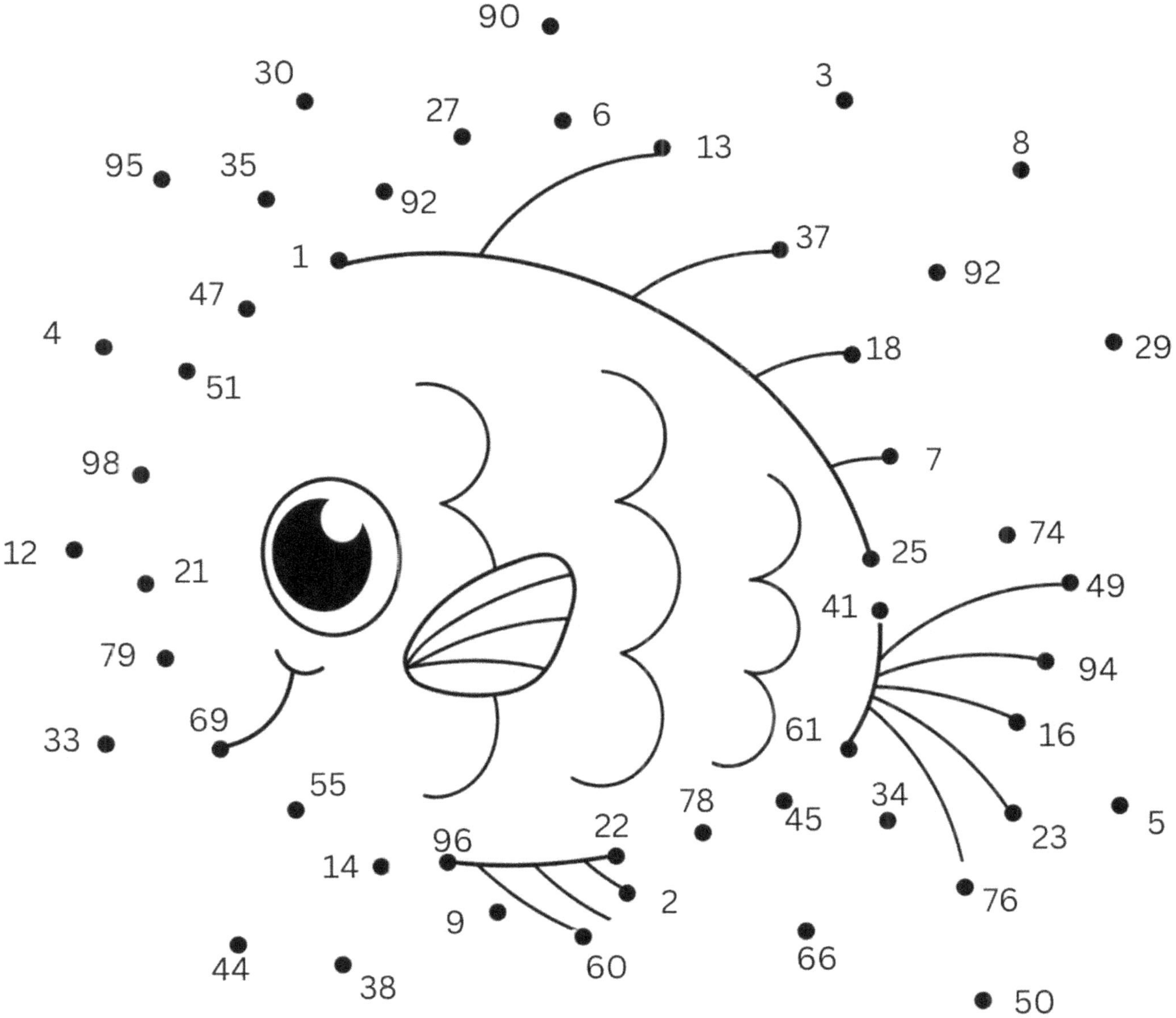
90
30
3
27
6
13
8
95
35
92
1
37
92
47
4
18
29
51
98
7
74
12
25
21
49
41
79
94
69
33
61
16
55
78
34
45
22
96
5
23
14
2
76
9
60
44
66
38
50

PUZZLE 16

A Joke for you

LETTER	ANSWER
A	3
B	4
C	7
D	9
E	12
F	14
G	16
H	18
I	24
J	27
K	30
L	31
M	33
N	35
O	38
P	40
Q	41
R	44
S	45
T	48
U	50
V	55
W	66
X	72
Y	77
Z	84

What did the big sausage say to the little sausage?

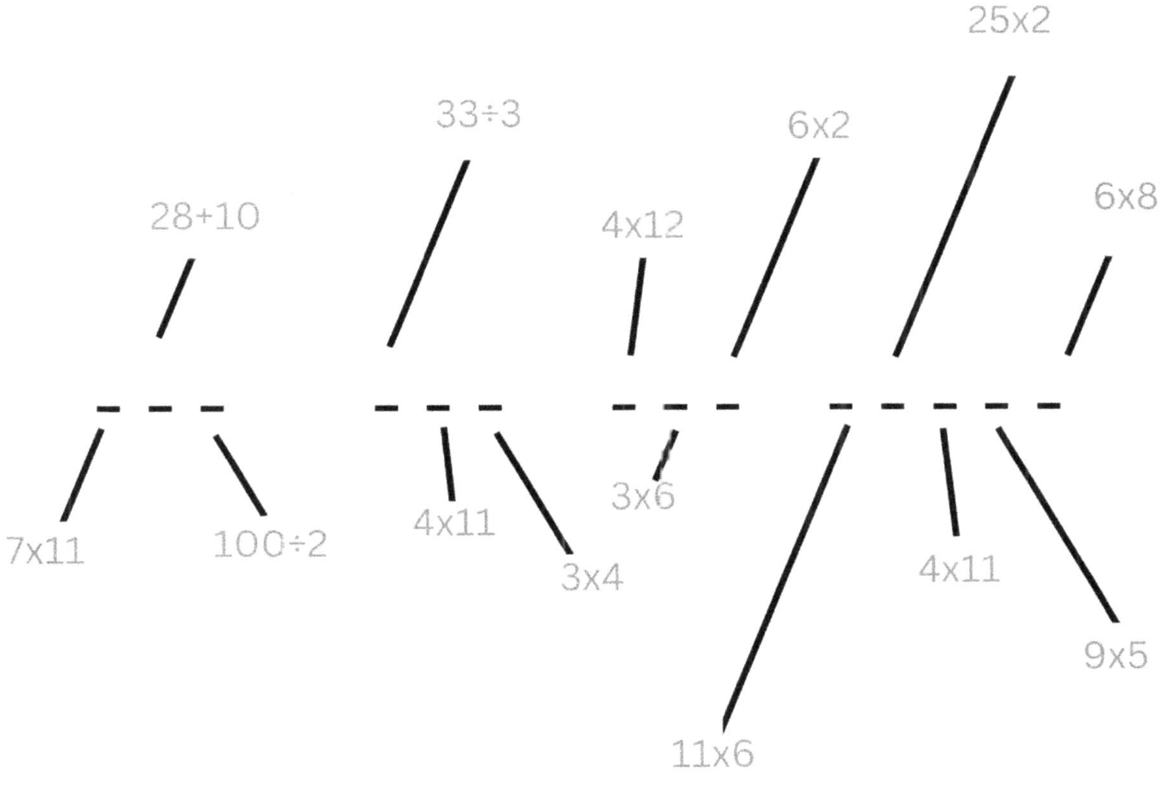

PUZZLE 17

A Joke for you

LETTER	ANSWER
A	3
B	4
C	7
D	9
E	12
F	14
G	16
H	18
I	24
J	27
K	30
L	31
M	33
N	35
O	38
P	40
Q	41
R	44
S	45
T	48
U	50
V	55
W	66
X	72
Y	77
Z	84

What do you call a sleeping cow?

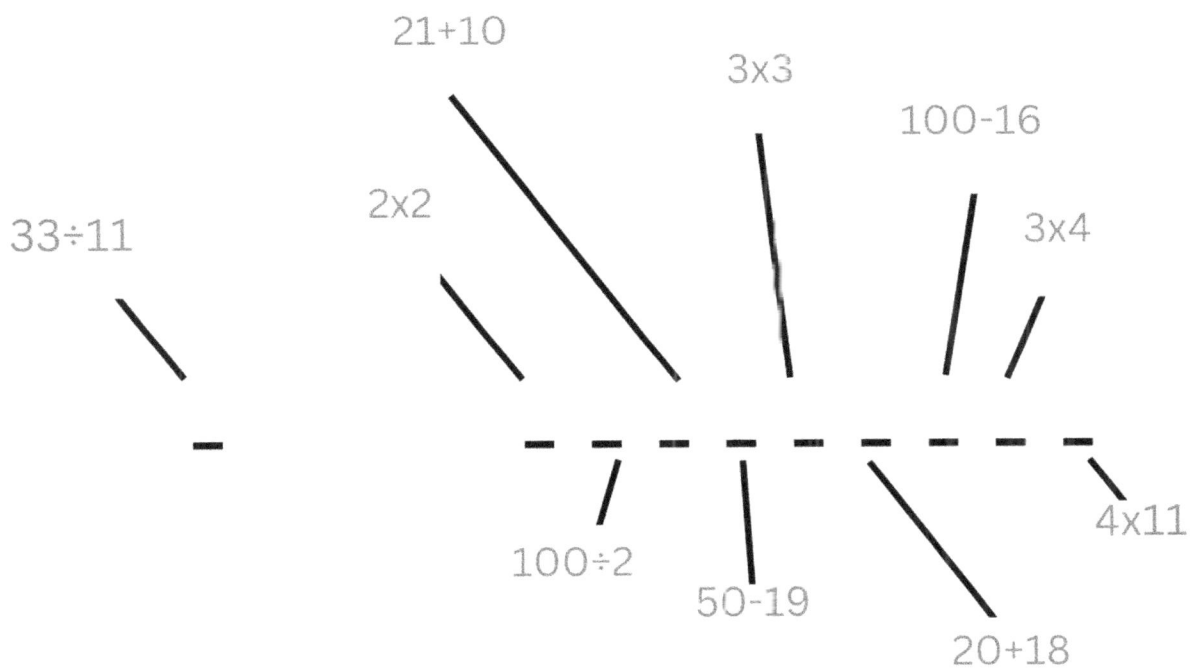

_   _ _ _ _ _ _ _ _ _

Blank	Clue
1	33÷11
2	2x2
3	100÷2
4	21+10
5	50-19
6	3x3
7	20+18
8	100-16
9	3x4
10	4x11

PUZZLE 18

Solve the equations and connect the dots in order

MATHS PROBLEM	ANSWER
start: 4 x 6 = 23 + ?	
3 x 7	
90 ÷ 9	
? - 22 = 3 x 9	
9 x 9	
? + 9 = 10 x 10	
61-2	
6 x 9	
27 + 30	
5 x 12	
40 -17	
36 ÷ 4	
8 + 9	
? - 2 = 9 x 5	
8 x 10	
? - 40 = 57	
48 ÷ 6	
2 x 7	
6 x 7	
8 x 11	
100 - 7	
? - 1 = 3 x 9	
4 + ? = 50	
22 + 30	
21 - 2	
? ÷ 4 = 6	
50 -7	
? - 4 = 49	
9 x 9	
100 - 4	
2 x 11	
? ÷ 11 = 6	

3
27
14
4
52
47
80
42
88
46
19
86
8
93
28
92
97
24
17
6
9
43
98
23
53
60
57
81
2
54
96
22
11
59
41
5
13
91
66
25
81
10
49
21
99
1

PUZZLE 19

Colour by Number

MATHS PROBLEM	ANSWER
BLACK	11
LIGHT GREEN	5
GREY	12
LIGHT BLUE	4
YELLOW	2
DARK GREEN	6
BROWN	10

2 x 6
20
÷
10
5+6
9+2
5+7
8-3
12-7
10÷2
20-9
11-7
20÷5
9+3
20-9
14-2
2X5
3+8
22÷2
100 ÷10
2X6
14-2
2x3
12÷2

PUZZLE 20

Solve the equations and connect the dots in order

MATHS PROBLEM	ANSWER
Start: 34 -33	
2 x 11	
7 x 11	
? ÷ 10 = 9	
27 ÷ 3	
21 + 20	
2  6 	
half of 100	
6  3 	
half of 14	
11 x 6	
102 - 5	
? x 3 = 33	
2 x 20	
16 ÷ 4	
30 x 2	
70 -12	
9 x 11	
88 ÷ 2	

49
43
2
206
94
54
22
9
90
41
50
7
3
1
77
630
66
88
97
362
11
40
52
4
.60
65
58
99
8
44

PUZZLE 21

Solve the equations and connect the dots in order

MATHS PROBLEM	ANSWER
start: 21 - 4	
6 x10	
51 - 4	
? - 10 = 19	
3 x 5	
10 x 7	
? ÷ 9 = 5	
88 - 20	
9 x 9	
? + 5 = 64	
23 + 10 + 10	
32 + 32	
80 - 9	
6 x 4	
? - 27 = 30	
12 + 29	
100 - 16	
2 x 11	
2 + 15 + 21	
61 - 5	
6 x 7 = ? - 32	
100 - 9	
8 x 10	
6 x 6	
71 - 10	
4 x 5	
26 + 16 + 10	
6 x 12	
45 ÷ 9	
79 - 30	
24 ÷ 8	
8 x 4	

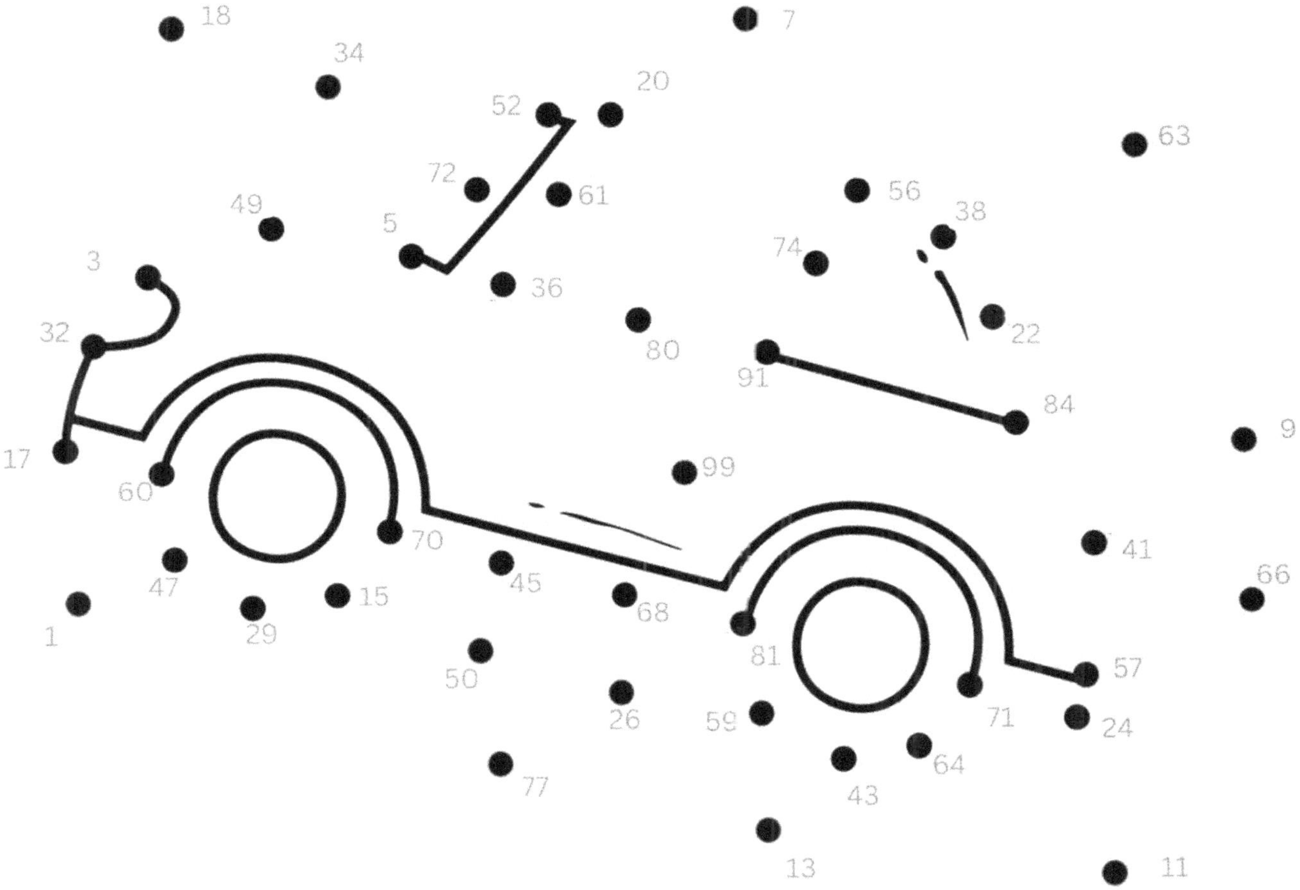
18
7
34
20
52
63
72
61
56
49
5
38
74
3
36
22
80
32
91
84
9
17
99
60
70
41
47
45
66
15
1
29
68
81
57
50
26
59
71
24
64
77
43
13
11

PUZZLE 22

Solve the equations and connect the dots in order

MATHS PROBLEM	ANSWER
start: 6 x 10	
7 x 5	
? x 10 = 130	
? ÷ 11 = 9	
41 x 2	
23 + 30	
? ÷ 2 = 13	
? + 7 = 5 x 10	
60 + 14	
4 x 4	
47 + 11	
4 x ? = 32	
12 x 4	
31 x 2	
8 x 10	
150 ÷ 10	
8 x 4	
13 + 10 + 29	
28 + 9	
5 x 4	
43 - ? = 6 x 7	
10 x 4	
21 + 10	
? x 10 = 590	
2 x 11	
100 ÷ 2	
6 x 12	
50 ÷ ? = 25	
60 ÷ ? = 30	
8 x 11	
7 x 6	
9 x 9	
40 + 17	
24 ÷ 6	
5 x 5	
3 x ? = 27	
45 ÷ ? = 9	
6 x 6	
36 ÷ 3	
? ÷ 11 = 4	
5 x 12	

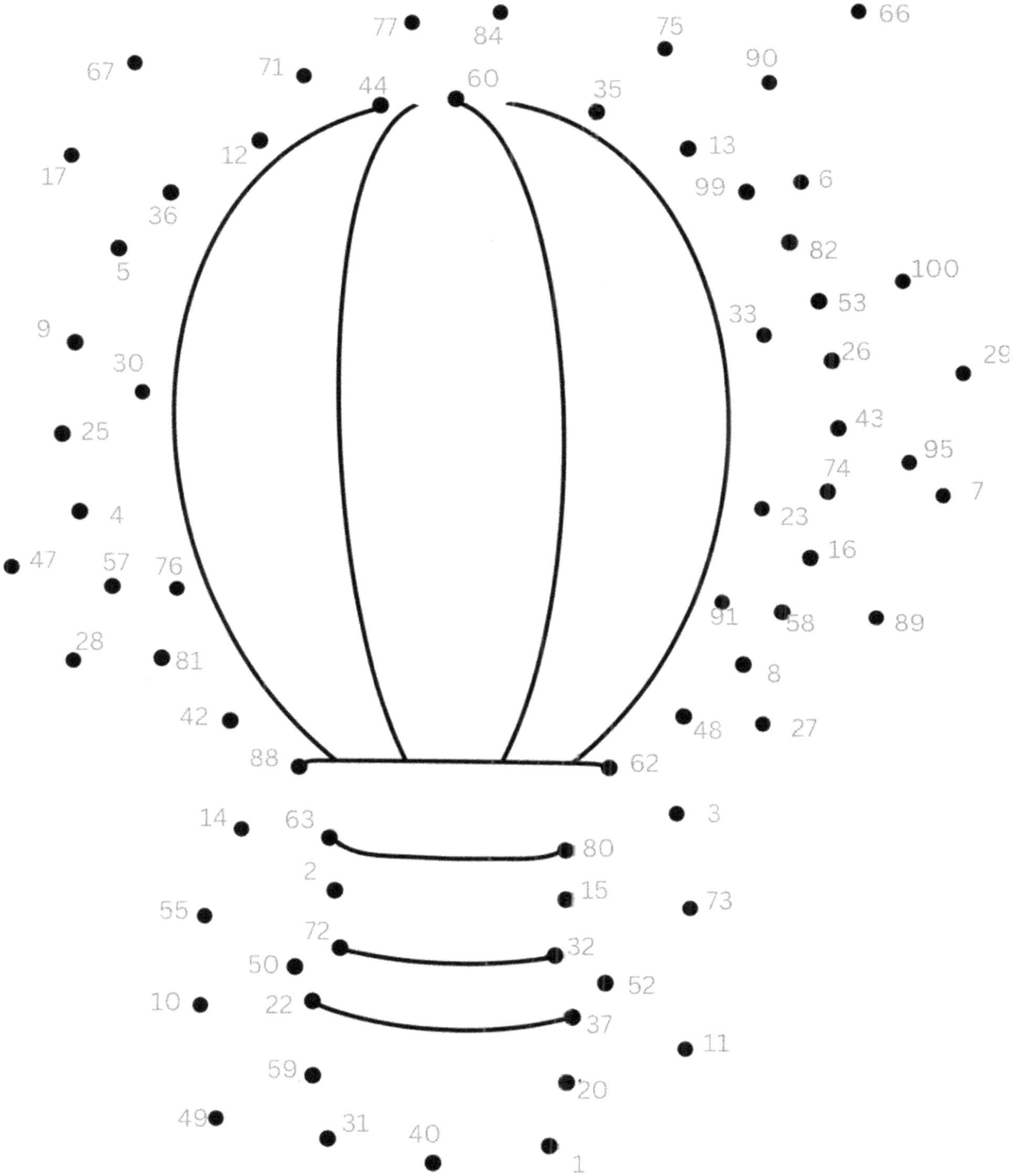
1
2
3
4
5
6
7
8
9
10
11
12
13
14
15
16
17
20
22
23
25
26
27
28
29
30
31
32
33
35
36
37
40
42
43
44
47
48
49
50
52
53
55
57
58
59
60
62
63
66
67
71
72
73
74
75
76
77
80
81
82
84
88
89
90
91
95
99
100

PUZZLE 23

Colour By Number

MATHS PROBLEM	ANSWER
Red	1, 13, 16, 200
Orange	2, 18, 24, 48
Grey	3, 14, 36, 50
Light blue	4, 9, 25, 32
Black	5, 21, 44, 66
Yellow	6, 12, 33, 72
Light Green	7, 10, 55, 60
Dark Green	8, 19, 30, 42
White	11, 20, 26, 40

2x2
50÷2
40-8
20+12
4x8
16÷4
30-5
12-8
9÷3
100÷2
2x10
6-2
9+16
5+6
2x7
6x6
3x11
3x4
44-40
8÷2
5x5
8x4
27÷3
3x3
5x5
8÷2
5x5
18÷2
3+9
2x3
8x2
2x9
4x12
6x4
80-8
3+1
12÷6
3x11
3x3
2x9
4x8
20+12
12÷2
4x8
18÷2
3x4
40-7
8-2
25-6
2x6
2x4
3x10
6x7
5x5
2x2
40÷2
13x2
5x6
22-3
4x5
7x6
6x5
100-60
21÷3
5x12
25+30
60-5
22÷2
10-3
30÷3
5x12
2x5
5x11
3+4

PUZZLE 24

A Joke for you

LETTER	ANSWER
A	3
B	4
C	7
D	9
E	12
F	14
G	16
H	18
I	24
J	27
K	30
L	31
M	33
N	35
O	38
P	40
Q	41
R	44
S	45
T	48
U	50
V	55
W	66
X	72
Y	77
Z	84

Why don't you play cards in the jungle?

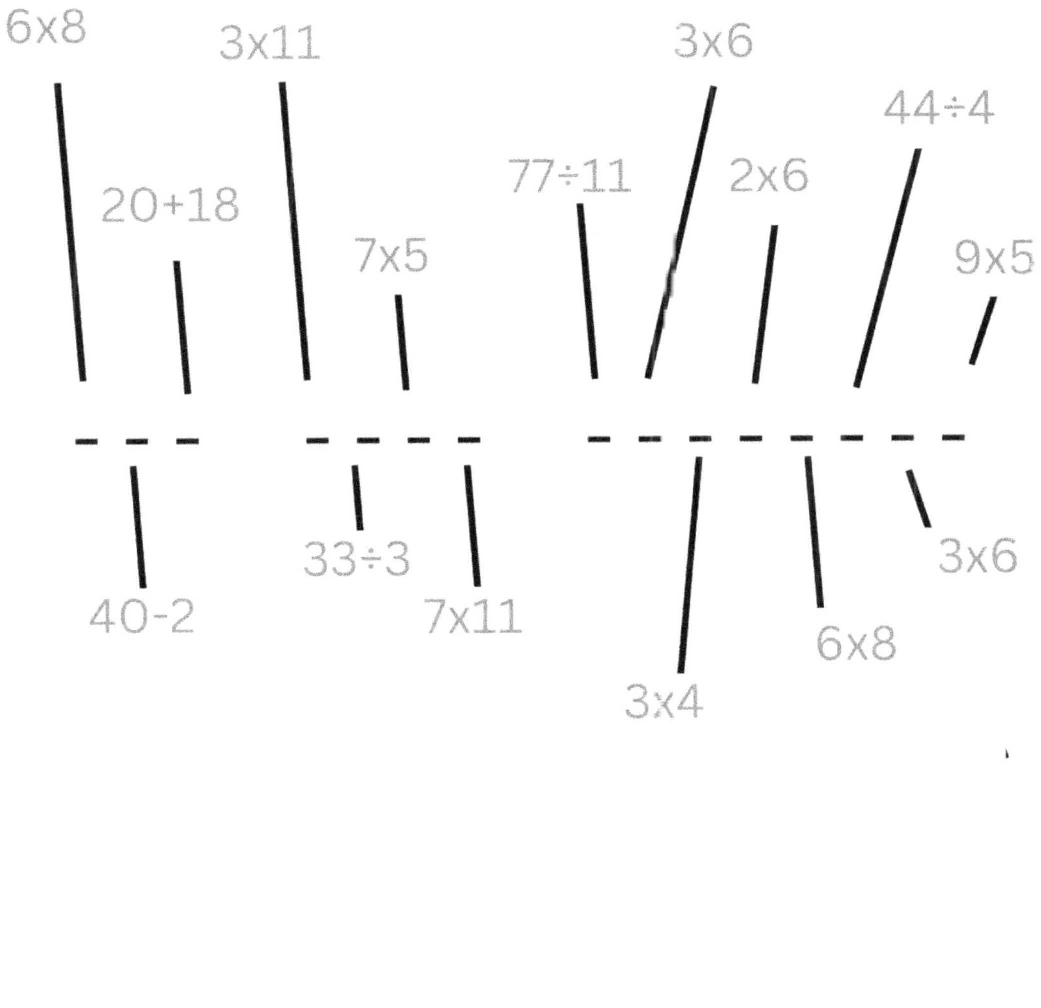

PUZZLE 25

Solve the equations and connect the dots in order

MATHS PROBLEM	ANSWER
start: 4 x 10	
3 x 11	
50 - 4	
6 x 4	
3 x 5	
? ÷ 5 = 7	
? x 6 = 18	
33 - 10	
? -1 = 6 x 6	
9 x 5	
6 x 5 = ? - 1	
5 x ? = 30	
38 + 30	
60 -2	
6 x 10	
90 - 8	
6 + 7	
20 + 30 + 11	
7 x 10	
21 + 20	
100 - 13	
60 - 1	
3 x 10	
80 ÷ ? = 8	
? x 12 = 48	
7 x 11	
32 - 3	
89 - 20	
6 x 7	
2 x 11	
46 - ? = 9 x 5	
7 x 7	
? x 10 = 110	
100 ÷ 5	
36 ÷ 4	
100 - 15	
? - 4 = 6 x 5	
60 ÷ 12	
? - 20 = 23	
? - 1 = 3 x 6	
3 x 7	
100 ÷ 2	
? x 6 = 12	
19 + 20	
4 x 4	
2 x 7	
13 x 2	
half of eighty	

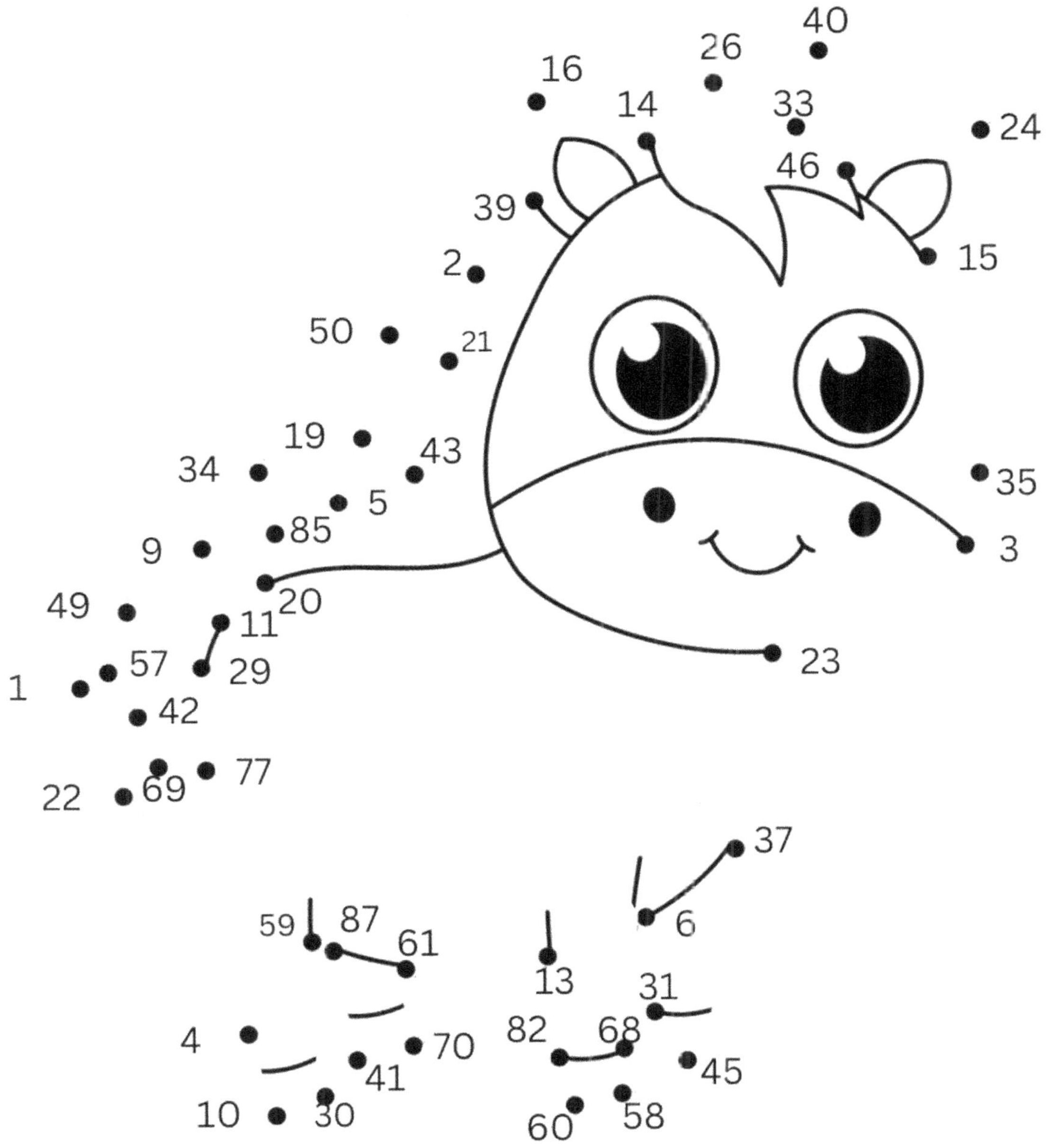
40
26
16
14
33
24
46
39
15
2
50
21
19
43
34
35
5
85
3
9
20
49
11
57
29
23
1
42
77
22
69
37
59
87
61
6
13
31
82
68
4
70
45
41
10
30
60
58

PUZZLE 26

Solve the equations and connect the dots in order

MATHS PROBLEM	ANSWER
start: 3 x 5	
a third of 9	
108 - 40	
261 - 200	
570 ÷ 10	
26 + 20	
3 x 11	
23 + 7 +13	
5 x 5	
? - 1 = 6 x 6	
10 + 7	
2 x 10	
3 x ? = 27	
40 - 11	
2 x 9	
6 x 7	
6 x 10	
2 x 4	
? ÷ 11 = 4	
21 + 10	
6 x ? = 24	
? ÷ 6 = 4	
147 - 100	
4 x 4	
? - 3 = 6 x 6	
16 + 10	
800 ÷ 10	
2 x 11	
520 ÷ 10	
32 + 30	
69 - 10	
18 ÷ 3	
365 - 300	
? ÷ 11 = 7	
20 - 7	
29 + 40	
101 - 30	

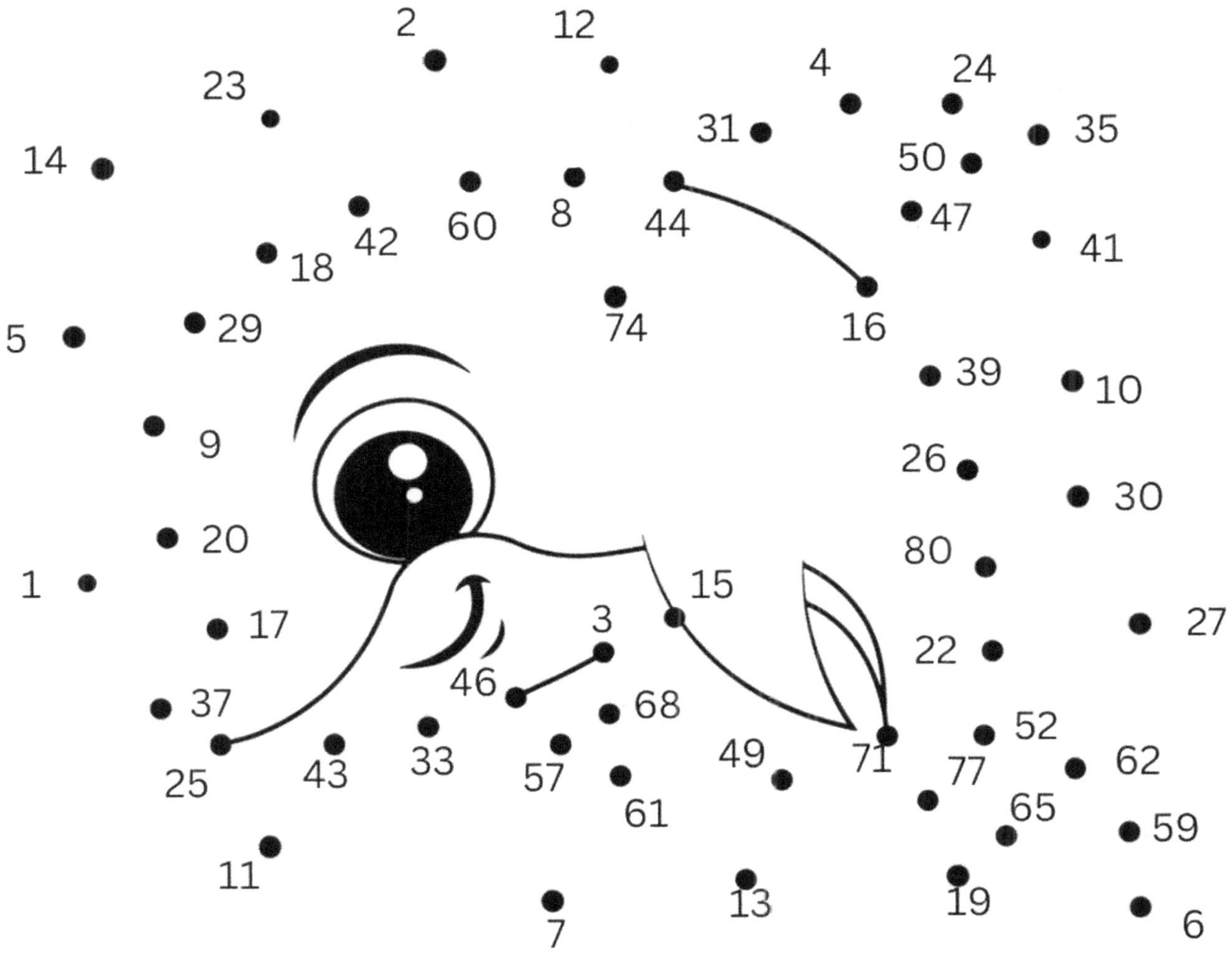
2
12
23
4
24
14
31
35
50
60
8
44
47
42
18
41
5
29
74
16
39
10
9
26
30
20
80
1
15
17
3
27
22
37
46
68
52
25
43
33
57
49
71
62
77
61
65
59
11
13
19
7
6

PUZZLE 27

Solve the equations and connect the dots in order

MATHS PROBLEM	ANSWER
start: 16 ÷ 4	
9 8	
790 ÷ 10	
6 x 5	
6 x 4	
5 x 12	
60 ÷ 12	
5 6	
13 + 10	
100 - 6	
38 + 30	
? + 3 = 6 x 10	
4 x 10	
? + 9 = 60	
? - 60 = 4	
5 x 10	
25 x 3	
290 ÷ 10	
9 x 11	
24 ÷ 12	
2 3 4	
44 ÷ 2	

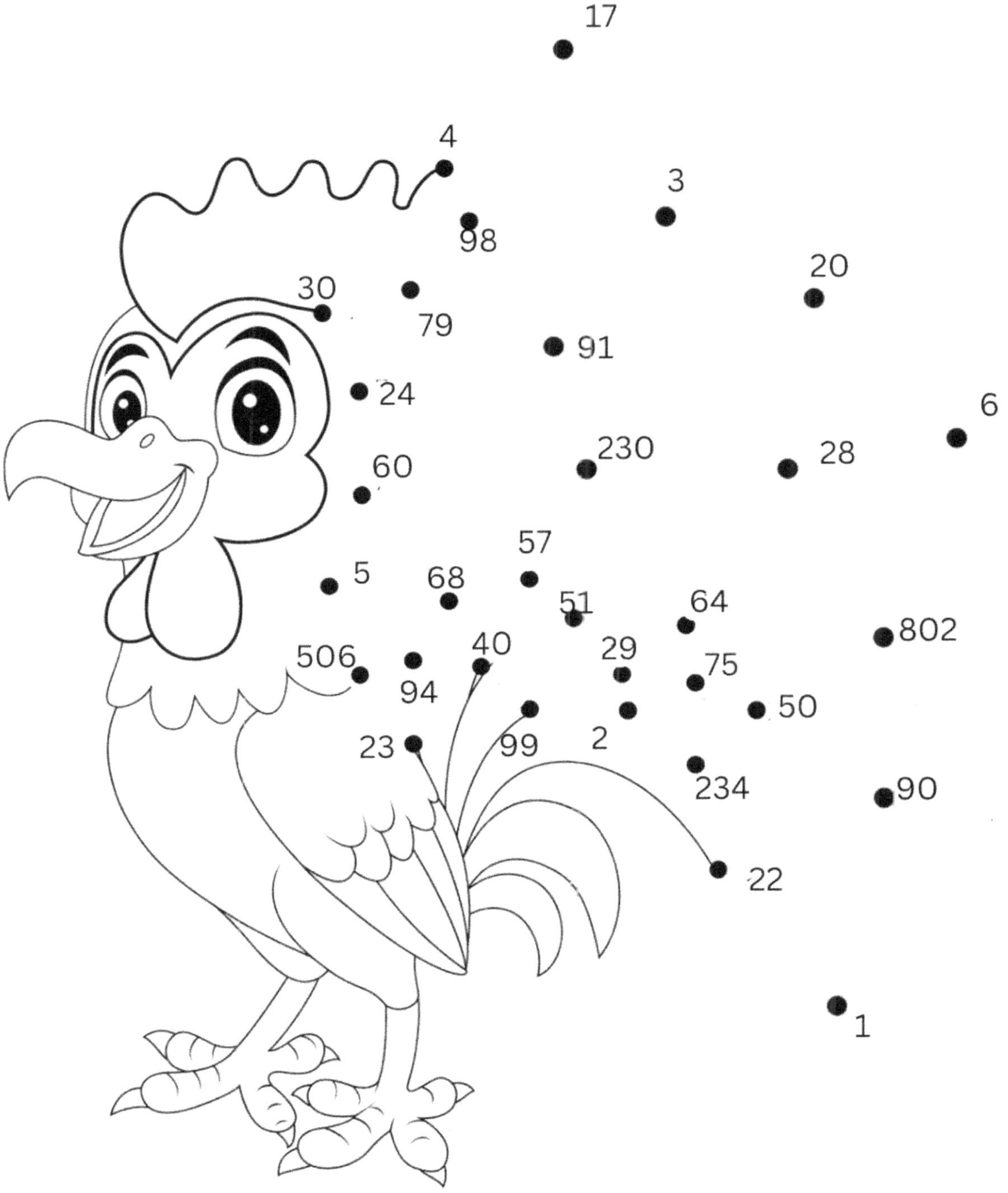
17
4
3
98
20
30
79
91
24
6
230
28
60
57
5
68
51
64
802
40
506
29
75
94
50
23
99
2
234
90
22
1

PUZZLE 28

Solve the equations and connect the dots in order

MATHS PROBLEM	ANSWER
start: 5 x 11	
3/4 of 100	
? - 2 = 6 x 6	
9 x 5	
6 x 7	
16 ÷ 2	
130 ÷ 10	
6 x 6	
18 ÷ 3	
9 x 6	
12 x 6	
55 ÷ 11	
30 ÷ 2	
21 + 20	
48 = ? x 12	
3 x 6	
? x 10 = 510	
28 +30	
24 ÷ 2	
2  5	
35 ÷ 5	
2  1 5 	
1/4 of 8	
1/2 of 28	
1/2 of 100	
7 x 11	
100 - 29	
170 ÷ 10	

2
215
10
3
17
14
7
71
55
502
250
12
38
1
75
77
50
58
33
51
45
18
432
42
72
41
4
54
13
6
8
5
15
36
11
81
509

PUZZLE 29

Colour By Number

MATHS PROBLEM	ANSWER
Red	1, 13, 16, 200
Green	2, 18, 24, 48
Grey	3, 14, 36, 50
Light blue	4, 9, 25, 32
Black	5, 21, 44, 66
Yellow	6, 10, 33, 72

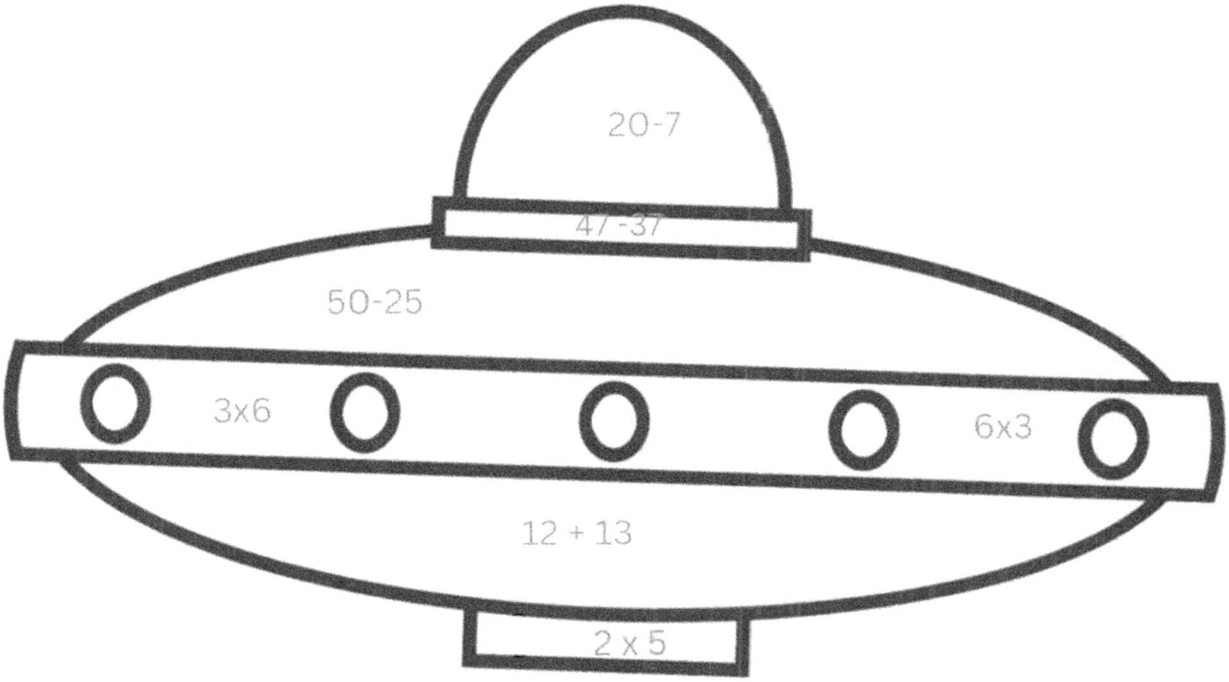
20-7
47-37
50-25
3x6
6x3
12 + 13
2 x 5

ANSWERS

PUZZLE 1

Solve the equations and connect the dots in order

MATHS PROBLEM	ANSWER
Start	1
25 x 2	50
1/2 of 50	25
100 ÷ ? = 50	2
40 - 13	27
2 x 5	10
100 - 51	49
42 - ? = 30	[illegible]
250 - 207	43
3 x 3	9
7 x 10	70
49 - [illegible]	29
3 x 5	15
400 ÷ 10	40
6 x 8	48
5 x 6	30
62 - 4	58
100 - 15	85
3 x ? = 21	7
20 + 30 + 23	73
10 + 41 [illegible]	51
5 x ? = 100	20
30 - 13	17
35 + 36	71

PUZZLE 2

Solve the equations and connect the dots in order

MATHS PROBLEM	ANSWER
5, ?, 25, 35	15
7 x 11	77
2 is 1/4 of ?	8
4 x4	16
50-13	37
30 x 3	90
60 -6	54
2 x 5	10
21, 26, ?, 36	31
100 -20	80
50 - 33	17
28 - 21	7
16 + 16	32
30 = ? - 61	91
41 + 30	71
26 ? 13	39
? x 9 = 27	3
6 x 11	66
100 -24	76
12 + 22	34
5 x 12	60

PUZZLE 3

Solve the equations and connect the dots in order

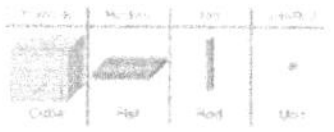

MATHS PROBLEM	ANSWER
5 x 12	60
60 -9	51
30 -11	19
25 + 12	37
4 3	430
3 2 1	321
6 x 11	66
44 - 11	33
2 x 11	22
90 - 13	77
21 -14	7
4 x 5	20
189 - 188	1
11 x 5	55
I have four 100's, two 10's and three singles	423
11 x 9	99
3 4	340
82 - 2	79
3 6	306

PUZZLE 4

Solve the equations and connect the dots in order

MATHS PROBLEM	ANSWER
[illegible]	[illegible]
[illegible]	37
2 9	209
[illegible]	71
[illegible]	28
[illegible]	59
[illegible]	95
[illegible]	3
[illegible]	53
40 + 12	52
[illegible]	44
1 7	17
[illegible]	26
[illegible]	25
[illegible]	75
[illegible]	78
[illegible]	60
3 4	340
[illegible]	7
[illegible]	38
[illegible]	30
[illegible]	60
2 [illegible]	205
[illegible]	40
[illegible]	30
20 + 20 + 1	41
[illegible]	35
[illegible]	17
[illegible]	2
[illegible]	[illegible]
[illegible]	51
7 x 11	77
[illegible]	47
[illegible]	31
7 x 6	42
[illegible]	99
6 x 12	72

PUZZLE 5

A Joke for you

It wasn't peeling well

PUZZLE 6

Solve the equations and connect the dots in order

MATHS PROBLEM	ANSWER
[illegible]	1
[illegible]	40
29 - 20	29
[illegible]	60
7 x 2	14
11 x 7	77
3 x 9	27
1 3	130
48 ÷ ? = 12	4
[illegible]	17
3 2	320
[illegible]	31
20 + 37	57
3 x 6	18
[illegible]	41
[illegible]	79
7 x ? = 35	7
13 x 2	26
2 3 4	234
40 - 12	38
8 + 11	19
6 x 4	24
40 + 3	43
2 6	206
[illegible]	56
3 x 11	33
3 x 3	9
2 x 11	22
4 x 5	20
4 x 11	44
1/2 of 100	50
[illegible]	81
41 + 20	61

PUZZLE 7

Colour by Number

MATHS PROBLEM	ANSWER
BLACK	1, 5, 10, 12
RED	13, 26, 50
GREY	2, 9, 14
YELLOW	7, 33, 100
LIGHT BLUE	3, 8, 24
ORANGE	6, 11,15
WHITE	4, 16 20, 70

PUZZLE 8

Colour by Number

MATHS PROBLEM	ANSWER
BLACK	11
RED	5
GREY	3
LIGHT BLUE	4
YELLOW	2
ORANGE	6

PUZZLE 9

Solve the equations and connect the dots in order

MATHS PROBLEM	ANSWER
[illegible]	[illegible]
[illegible]	1
[illegible]	[illegible]
[illegible]	[illegible]
[illegible]	[illegible]
[illegible]	[illegible]
[illegible]	6
4 2	[illegible]
[illegible]	[illegible]
[illegible]	[illegible]
[illegible]	[illegible]
3 x 3	9
[illegible]	[illegible]
[illegible]	[illegible]
[illegible]	[illegible]
[illegible]	[illegible]
[illegible]	[illegible]
[illegible]	5
[illegible]	[illegible]
[illegible]	4
[illegible]	[illegible]
[illegible]	[illegible]
[illegible]	[illegible]
[illegible]	[illegible]
[illegible]	[illegible]
[illegible]	[illegible]
[illegible]	[illegible]
[illegible]	6
[illegible]	[illegible]
[illegible]	[illegible]
[illegible]	[illegible]
[illegible]	3
[illegible]	[illegible]
[illegible]	[illegible]
[illegible]	[illegible]
[illegible]	2

PUZZLE 10

Solve the equations and connect the dots in order

MATHS PROBLEM	ANSWER
start: 21 + 40	61
100 - 18	82
25 x 4	100
3 4	340
4 x 11	44
50 - ? = 1	49
29 + 10	39
30 - 13	17
30 x 3	90
1 3 6	136
3 x 10	30
3 x 5	15
4 x 20	80
5 x 9	45
2 1 2	212
6 x 3	18
10 x 4	40
5 x 5	25
35 x 2	70
4 6	460
50 - 9	41
16 x 2	32
5 x 2	10
100 - 39	61

PUZZLE 11

Solve the equations and connect the dots in order

MATHS PROBLEM	ANSWER
1/4 of 40	10
40 - 11	29
60 - 8	52
6 3	603
6 x 12	72
11 x 8	88
60 -1	59
6 x 4	24
3 2 7	327
5 X 9	45
25 +26	51
8 + 11	19
30 x 2	60
half of 30	15
3 x 30	90
? ÷ 3 = 9	27
9 x 11	99
3 x 11	33
8 x 5	40
50 + 32	82
5 2 4	524
2 x 11	22
10 + 3	13
2 3 1	231

PUZZLE 12

A Joke for you

Moosli

PUZZLE 13

Colour by Number

MATHS PROBLEM	ANSWER
BLACK	1, 7, 42
GREEN	5, 11, 24
GREY	3, 8, 32
LIGHT BLUE	4, 27, 50
YELLOW	2, 9, 44
ORANGE	6, 23, 56
WHITE	10,14, 25

PUZZLE 14

Solve the equations and connect the dots in order

MATHS PROBLEM	ANSWER
Start: 46 = 26 + 19 + ?	1
30 x 2	60
5 x3	15
? ÷ 3 = 10	30
22 + 30	52
30 - 13	17
9 x 11	99
20 + 50	70
3 x 6	18
32 + 27	59
33 - 22	11
33 + 36	69
100 - 9	91
25 + 36	61
33 + 20	53
22 - 2	19
? x 5 = 30	6
7 x 10	70
6 x 12	72
108 - 16	92
47 - 33	14
126 - 40	86
9 x ? = 45	5
100 -19	81
80 ÷ ? = 8	10
5 + 8	13
100 - 49	51
70 - 35	35
2 x 10	20

PUZZLE 15

Solve the equations and connect the dots in order

MATHS PROBLEM	ANSWER
start: 46 - ? = 9 x 5	1
42 + 50	92
9 x 3	27
72 ÷ 12	6
? x 2 = 26	13
50 - 13	37
3 x 6	18
35 ÷ ? = 5	7
1/4 of 100	25
? - 20 = 21	41
100 - 26	74
23 + 26	49
104 - 10	94
4 x 4	16
56 - 43	23
46 + 30	76
14 + 20	34
21 + 40	61
9 x 5	45
84 - 6	78
11 x 2	22
24 ÷ 12	2
5 x 12	60
3 x ? = 27	9
100 - 4	96
2 x 7	14
11 x 5	55
75 - 6	69
80 - 30 + 29	79
? ÷ 7 = 3	21
104 - 6	98
65 - 14	51
20 + 20 + 7	47
400 -399	1

PUZZLE 16

A Joke for you

you are the wurst

PUZZLE 17

A Joke for you

a bulldozer

PUZZLE 18

Solve the equations and connect the dots in order

MATHS PROBLEM	ANSWER
start: 4 x 6 = 23 + ?	1
3 x 7	21
90 ÷ 9	10
? - 22 = 3 x 9	49
9 x 9	81
? + 9 = 10 x 10	91
64 - 9	55
6 x 9	54
27 + 30	57
5 x 12	60
40 - 17	23
36 ÷ 4	9
8 + 9	17
? - 2 = 9 x 5	47
8 x 10	80
? - 40 = 57	97
48 ÷ 6	8
2 x 7	14
6 x 7	42
8 x 11	88
100 - 7	93
? - 1 = 3 x 9	28
4 + ? = 50	46
22 + 30	52
21 - 3	18
? ÷ 4 = 6	24
50 - 7	43
? - 4 = 49	53
9 x 9	81
100 - 4	96
2 x 11	22
? ÷ 11 = 6	66

PUZZLE 19

Colour by Number

MATHS PROBLEM	ANSWER
BLACK	11
LIGHT GREEN	5
GREY	12
LIGHT BLUE	4
YELLOW	7
DARK GREEN	6
BROWN	10

PUZZLE 20

Solve the equations and connect the dots in order

MATHS PROBLEM	ANSWER
Start: 34 -33	1
2 x 11	22
7 x 11	77
? ÷ 10 = 9	90
27 ÷ 3	9
21 + 20	41
2 [illegible] 6	206
half of 100	50
6 [illegible] 3	630
half of 14	7
11 x 6	66
102 - 5	97
? x 3 = 33	11
2 x 20	40
16 ÷ 4	4
30 x 2	60
70 -12	58
9 x 11	99
88 ÷ 2	44

PUZZLE 21

Solve the equations and connect the dots in order

MATHS PROBLEM	ANSWER
start: 21 - 4	17
6 x10	60
51 - 4	47
? - 10 = 19	29
3 x 5	15
10 x 7	70
? ÷ 9 = 5	45
88 - 20	68
9 x 9	81
? + 5 = 64	59
23 + 10 + 10	43
32 + 32	64
80 - 9	71
6 x 4	24
? - 27 = 30	57
12 + 29	41
100 - 16	84
2 x 11	22
? + 15 + 21	38
61 - 5	56
6 x 7 = ? - 32	74
100 - 9	91
8 x 10	80
6 x 6	36
71 - 10	61
4 x 5	20
26 + 16 + 10	52
6 x 12	72
45 ÷ 9	5
79 - 30	49
24 ÷ 8	3
8 x 4	32

PUZZLE 22

Solve the equations and connect the dots in order

MATHS PROBLEM	ANSWER
start: 6 x 10	60
7 x 8	56
? x 10 = 120	12
? ÷ 11 = 9	99
41 - ?	37
23 + 30	53
? ÷ 2 = 13	26
? - 7 = 3 x 12	43
60 + 14	74
4 x 4	16
47 + 11	58
? + 7 = 15	8
12 x 4	48
31 x 2	62
8 x 10	80
150 ÷ 10	15
8 x 4	32
13 + 18 + 26	57
28 + 9	37
5 x 6	30
43 - ? = 6 x 7	1
10 x 4	40
21 + 10	31
? x 20 = 560	59
2 x 11	22
100 ÷ 2	50
6 x 12	72
40 - ? = 38	2
90 ÷ ? = 30	63
8 x 11	88
7 x 6	42
9 x 9	81
40 + 17	57
24 ÷ 6	4
5 x 5	25
3 x ? = 27	9
45 ÷ ? = 9	5
6 x 6	36
36 ÷ 3	12
? ÷ 11 = 4	44
5 x 12	60

PUZZLE 23

Colour By Number

MATHS PROBLEM	ANSWER
Red	1, 13, 16, [illegible]
Orange	2, 18, 24, 44
Grey	3, 14, 36, [illegible]
Light blue	4, 9, 25, [illegible]
Black	5, 21, 44, [illegible]
Yellow	6, 12, 33, [illegible]
Light Green	7, 10, 55, [illegible]
Dark Green	8, 19, 30, [illegible]
White	11, 20, 26, [illegible]

PUZZLE 24

A Joke for you

too many cheetahs

PUZZLE 25

Solve the equations and connect the dots in order

MATHS PROBLEM	ANSWER
start: 4 x 10	40
3 x 11	33
[illegible]	48
6 x 4	24
3 x 5	15
[illegible]	33
? x 4 = 16	2
38 - 16	22
? - 1 = 6 x 6	37
9 x 5	45
[illegible]	[illegible]
[illegible]	5
[illegible]	60
[illegible]	54
[illegible]	60
[illegible]	87
[illegible]	11
[illegible]	6
[illegible]	[illegible]
[illegible]	[illegible]
[illegible]	[illegible]
[illegible]	54
[illegible]	[illegible]
[illegible]	13
[illegible]	4
7 x 11	77
72 ÷ 3	29
99 - 30	69
6 x 7	42
2 x 11	21
[illegible]	1
7 x 7	49
? x 10 = 120	12
100 ÷ 5	20
28 + 3	9
150 ÷ 15	10
[illegible]	34
60 ÷ 12	5
? + 23 = 22	45
[illegible]	11
3 x 7	21
100 ÷ 2	50
? x 4 = 12	3
98 ÷ 26	37
4 x 4	16
2 x 7	14
12 x 2	24
[illegible]	43

PUZZLE 26

Solve the equations and connect the dots in order

MATHS PROBLEM	ANSWER
start: 3 x 5	15
a third of 9	3
108 - 40	68
261 - 200	61
570 ÷ 10	57
26 + 20	46
3 x 11	33
[illegible]	43
5 x 5	25
? - 1 = 6 x 6	37
10 + 7	17
2 x 10	20
3 x ? = 27	9
40 - 11	29
2 x 9	18
6 x 7	42
6 x 10	60
2 x 4	8
? ÷ 11 = 4	44
21 + 10	31
6 x ? = 24	4
? ÷ 6 = 4	24
147 - 100	47
4 x 4	16
? - 3 = 6 x 6	39
16 + 10	26
800 ÷ 10	80
2 x 11	22
520 ÷ 10	52
32 + 30	62
69 - 10	59
18 ÷ 3	6
365 - 300	65
? ÷ 11 = 7	77
20 - 7	13
25 + 40	49
101 - 30	71

PUZZLE 27

Solve the equations and connect the dots in order

MATHS PROBLEM	ANSWER
start: 16 ÷ 4	4
9 8	98
790 ÷ 10	79
6 x 5	30
6 x 4	24
5 x 12	60
60 ÷ 12	5
5 6	506
13 + 10	23
100 - 6	94
38 + 30	68
? + 3 = 6 x 10	57
4 x 10	40
? + 9 = 60	51
? - 60 = 4	64
5 x 10	50
25 x 3	75
290 ÷ 10	29
9 x 11	99
24 ÷ 12	2
2 3 4	234
44 ÷ 2	22

PUZZLE 28

Solve the equations and connect the dots in order

MATHS PROBLEM	ANSWER
start: 5 x 11	55
3/4 of 100	75
? - 2 = 6 x 6	38
9 x 5	45
6 x 7	42
16 ÷ 2	8
130 ÷ 10	13
6 x 6	36
18 ÷ 3	6
9 x 6	54
12 x 6	72
55 ÷ 11	5
30 ÷ 2	15
21 + 20	41
48 = ? x 12	4
3 x 6	18
? x 10 = 510	51
28 + 30	58
24 ÷ 2	12
2 5	250
35 ÷ 5	7
2 1 5	215
1/4 of 8	2
1/2 of 28	14
1/2 of 100	50
7 x 11	77
100 - 29	71
170 ÷ 10	17

PUZZLE 29

Colour By Number

MATHS PROBLEM	ANSWER
Red	1, 13, 16, 200
Green	2, 18, 24, 48
Grey	3, 14, 36, 50
Light blue	4, 9, 25, 32
Black	5, 21, 44, 66
Yellow	6, 10, 33, 72

Dear Readers,

Thank you for taking the time to read our book. As a small independent company your support immensely helps us and encourages us to pursue new work. We would be grateful if you can leave a review on Amazon, Goodreads or any other forum that you think would be helpful in spreading the word and helping other readers decide whether or not to read the book. We are always excited to hear from anyone who would like to collaborate on a book. Don't forget to email us any feedback at *learningthroughfun1@gmail.com*.

Publications in this Range

Learning Through Fun - Maths Dot to Dot & Other Fun Activities Years 2 -3

Learning Through Fun - Maths Dot to Dot & Other Fun Activities Years 3 -4

Learning Through Fun - Maths Cooking Year 2

Learning Through Fun - Maths Cooking Year 3

Learning Through Fun - Maths Cooking Year 4

Learning Through Fun - Maths Cooking Year 5

www.ingramcontent.com/pod-product-compliance
Lightning Source LLC
Chambersburg PA
CBHW081748200326
41597CB00024B/4433